Introduction
to
Difference
Equations

Introduction
to
Difference
Equations

with illustrative examples from
Economics, Psychology, and Sociology

SAMUEL GOLDBERG

Associate Professor of Mathematics
Oberlin College

John Wiley & Sons, Inc.   New York • London • Sydney

10 9 8

© 1958 by JOHN WILEY & SONS, INC.

PRINTED IN THE UNITED STATES OF AMERICA

Library of Congress Catalog Card Number: 58-10223

ISBN 0 471 31051 4

To
Marcia

Preface

This volume is a much expanded and entirely revised version of a short monograph on difference equations originally written in 1954 at the invitation of the Social Science Research Council Committee on the Mathematical Training of Social Scientists. Although I hope it will find favor with students and teachers of mathematics, this book is primarily intended for social scientists who wish to understand the basic ideas and techniques involved in setting up and solving difference equations. Wherever possible, the mathematical topics in the text are related to and illustrated by material drawn from the social sciences, especially economics, psychology, and sociology. Problems for solution, as well as illustrative examples throughout the text, often have their source in research papers and books in these fields. References to this literature are given in footnotes.

Some facility with standard algebraic techniques and an acquaintance with the essentials of trigonometry, together with at least a modicum of that elusive quality known as mathematical maturity, should be adequate preparation for reading this book. Such topics as the summation notation, proof by mathematical induction, the binomial formula, determinants, polar form of complex numbers, and de Moivre's theorem—which some readers may recall only faintly— are briefly reviewed as they are needed in the text. Only four sections treat special topics requiring some knowledge of material ordinarily taught in a college course in calculus. These sections (three of which explore the striking similarities between difference and differential equations, and the fourth involves infinite series) are starred and are neither used nor referred to in the remainder of the text. Of course, students whose mathematical training does not extend beyond algebra, trigonometry, or even a first course in calculus cannot expect to be introduced to more than a small part of the extensive literature on difference equations. Happily, however, this small part is quite important, for it is the foundation upon which the more advanced mathematical theory is built, and it is very widely applied in many fields, including the social sciences.

In the short introductory Chapter 0 we look at two examples: the experimental extinction of a learned response, and a multiplier-accelerator interaction process in economic dynamics. Although treated in greater detail later on, they are introduced here to illustrate how difference equations (and the need for mathematical methods to explore the implications of these equations) arise in the context of social science problems.

Those parts of the calculus of finite differences that are essential to the main theme of this book are carefully developed in Chapter 1. The finite difference operators are defined here and their important properties are elaborated. The function concept is introduced early and used to the fullest in order to present this material in a way consistent with modern mathematical usage. Readers who prefer to meet difference equations early in their study program could proceed immediately to Chapter 2, referring back to Chapter 1 as the need arises.

Difference equations are introduced in Chapter 2, where a complete treatment of the simplest first-order difference equation with constant coefficients is given. Since the solution of a difference equation is a sequence, we define sequences of real numbers and carefully introduce the ideas of limit and convergence. It is then possible to discuss the variety of limiting behaviors exhibited by solutions of even this very simple difference equation. There follow some applications of this equation in the social sciences: compound interest and amortization of debts, the classical Harrod-Domar-Hicks model for growth of national income, Metzler's pure inventory cycle, the Bush-Mosteller probability model for simple learning, the Bernoulli utility-wealth relation, and the Weber-Fechner law.

Chapter 3 treats the linear difference equation with constant coefficients. For simplicity, the case of order 2 is discussed in detail (including the oft-neglected idea of a fundamental set of solutions) and the methods developed are later carried over to the general case of order n. Again the important question of limiting behavior of solutions is discussed and applied to a variety of social science examples.

Chapter 4 consists of five sections, three of which, on equilibrium values and stability of difference equations, first-order equations and cobweb cycles, and a boundary-value (or characteristic-value) problem, are quite brief. One of the two fairly long concluding sections treats generating functions (with special emphasis on their use in studying ordinary and partial difference equations); the other develops matrix methods for the solution of systems of simultaneous difference equations. Although these topics are separately treated and in some ways unrelated, their inclusion in this final chapter is due to the fact that they share the following properties: (1) they require no greater mathematical background than assumed of readers of this book, (2) they are presently being used in the social science literature, and (3) aside from their many applications, they are of intrinsic mathematical worth and interest. In addition, the sections on generating functions and matrix methods open the way for the inclusion of some introductory material from the field of probability, a mathematical theory of increasing importance in the social sciences. As usual, the line had to be drawn somewhere; limitations of space prevented the treatment of still other topics, such as finite difference interpolation formulas, functional equations, and approximate solutions of certain nonlinear difference equations. A selected list of references for further study follows Chapter 4.

Problems for solution are intended to supplement the worked-out illustrative examples in the text and to enable the reader to check his understanding of new definitions, theorems, and methods. A brief hint or even a concise outline of the procedure to follow in solving the problem is often given. Answers to a representative selection of about 115 problems (out of a total of over 250) are collected in a separate section at the end of the book. As mentioned earlier, many problems are designed to illustrate the application of finite differences and difference equations in the social sciences. Although I have tried to select applications that would interest the social scientist, I have made no attempt to evaluate the worth of these applications. This seems to me rather in the domain of those whose competence in the applied fields is considerably greater than mine.

Throughout, stress is laid on the explanation of fundamental concepts and patterns of mathematical reasoning, rather than merely on techniques of problem-solving. For didactic reasons, it was often felt desirable to develop mathematical methods in the context of a particular application, to repeat basic concepts for emphasis, to consider simplified models at the outset, and to generalize (if possible at the level of this book) only after a thorough analysis of the simple situation had been made. For example, many of the important ideas and methods associated with boundary-value problems are introduced in Section 4.3 by means of an exhaustive treatment of a fairly simple particular example. It did not seem appropriate to go into the general theory in this volume, but the value of having studied this section will be apparent to a reader pursuing the subject in a more advanced work. Similarly, a good deal of the material on matrix methods in Section 4.5 is presented in the context of some problems leading to a system of two simultaneous difference equations. The consequent limitation to 2×2 matrices makes the material easy to grasp; nevertheless it allows immediate extension to more general systems. I am hopeful that such procedures will make this book more appealing and understandable to the mathematically oriented social scientist.

Students of mathematics will also find this volume of some interest, for two reasons. First, this introduction to difference equations may be of value in promoting an increased awareness on the part of such students for the applications of mathematics in the social sciences. The second and more weighty reason stems from the fact that some very important mathematical ideas (such as function, operator, convergence, fundamental set of solutions, generating function transformation, characteristic value, and matrix) are carefully introduced here in a fairly simple and intuitive context. It should be helpful for a student to learn about them first in this form, without the analytic difficulties usually accompanying them in later, more advanced courses.

It is with genuine appreciation that I here express my thanks to the following individuals who have read one or another of the preliminary versions of the manuscript and who were kind enough to send me their criticisms and suggestions for improvement: R. R. Bush, C. Christ, W. G. Madow, G. A. Miller, F. Mosteller, H. Raiffa, P. A. Samuelson, R. M. Solow, G. L. Thompson, and R. M. Thrall.

Comments from readers are always welcome.

SAMUEL GOLDBERG

Oberlin, Ohio
March 1958

Contents

0

Introduction

Problems in which some variable may conveniently be assumed to have only a discrete set of possible values often lead to mathematical models involving difference equations. In economics, for example, such a variable is time. The values of important economic quantities (income, savings, consumption, etc.) are ordinarily available at certain uniformly spaced time intervals. Data may be accumulated each month, quarter, year, or even each 10 years, as in the national census. All quantities are dated, each with the time period for which it applies. Thus we may speak of the national income at some initial time, which we choose to denote by $t = 0$, and then at the end of the next time period, at $t = 1$, and then at $t = 2$, etc. The economist, in what is called *period analysis*, studies the behavior of national income, and other economic variables, over this discrete set of time values. We shall see that an important ingredient of this analysis is the difference equation.

As a second example, consider free-recall verbal learning experiments in which a list of words is read aloud to a subject who is then asked to write down all the words he can recall. This constitutes one trial of the total experiment. Then the order of the words is randomized and the procedure is repeated, thus completing the second trial. At each trial we may compute the proportion of words recalled by the subject. The experiment, continued until this proportion is approximately equal to its ultimate stable value, yields the values of the proportion of words recalled in trial number 1, in trial number 2, etc. The psychologist is interested in studying the properties of this sequence of values. In this case, the discrete variable is the trial number and once again a difference equation may be expected to be a helpful mathematical tool.

In panel surveys in sociological investigations, a fixed group of people (the panel) may be asked the same set of questions at periodic time intervals, say at the end of each of the 6 months preceding an election. Here we have a first poll, then a second, a third, etc., and one studies a person's responses as a function of the poll number.   The existence of this discrete variable, assuming the integer values 1 to 6 in the example cited, allows the use of a difference equation model for the study of such data.

A large number of illustrative examples of this kind, all leading to a mathematical analysis involving difference equations, will be given in this book.   Difference equations have appeared, and continue to appear, in the literature of the social and behavioral sciences.   Before developing the mathematical theory of such equations, it might be well to consider one or two examples in greater detail.   Our only aim here is to point out, without regard for mathematical niceties, how a difference equation can arise in connection with problems of the social sciences.   The crucial problems of justifying the assumptions to be made and of testing the usefulness of this particular mode of analysis in the social sciences are not within our purview.   These are problems for the social scientist, not the mathematician.

Example 1

Let us suppose that a subject is introduced into the following over-simplified learning situation: (1) a stimulus is presented, (2) the subject may or may not react to this stimulus, but (3) if he does respond positively, he is by some means discouraged from repeating this response.   To fix ideas, consider the example in which a rat, previously perfectly conditioned to running a straight runway to find food at its end, is placed in the starting box.   In a specified subsequent time interval (long enough to permit completion of the run, but not long enough to allow dawdling along the way) the rat either makes the complete run or does not.   If he does, he is disappointed to find that the food reward is no longer present.   Let us call the completion of the run in the allotted time a positive response. After this trial run, the rat is once again placed in the starting box and another trial takes place.   In this way the rat is subjected to many repeated trial runs, each of which may or may not result in a positive response.   If we imagine a large number of rats similarly, but independently, used as subjects in these repeated runway trials, then we can compute the proportion of rats responding positively in trial number 1, in trial number 2, etc.   Intuitively, we expect that the response of running to the end of the runway in the fixed allotted time interval will be extinguished owing to the absence of reward and that this "learning" will manifest itself in an ever-decreasing proportion of the rats who respond positively.

Now it actually turns out to be more convenient for the mathematical model to study a subject's *probability* of making a positive response, rather than the *proportion* of positive responses in a large group of subjects.   (Of course, when using this probability model one ordinarily takes this empirical proportion as an estimate of the theoretical probability. The reader may, if he chooses, think of proportion, expressed in decimal form, rather than probability.) This probability is 0 if the response is never made, is 1 if the stimulus always elicits a positive response, and generally is some number between these extreme values. Since the subject learns as the experimental trials are run, the probability of a positive response will change from trial to trial. So we let n assume the integral values $0, 1, 2, \cdots$ (to represent the trial number) and then define p_n as the probability of a positive response in trial number n. We thus have a sequence of probabilities $p_0, p_1, p_2, \cdots$ describing the behavior of subjects in repeated trials of the kind delineated.

A subject's behavior will depend on his conditioning before the experiment. The symbol p_0 denotes the initial probability of a positive response before the first trial. Thus $p_0 = 1$ would be interpreted as perfect conditioning toward the positive response; a subject equally disposed toward positive and negative responses would be assigned the value $p_0 = 0.5$.

If the probabilities change from trial to trial, we must specify the way in which they vary. Since we want the probabilities to decrease as the subject learns, we require that $(p_{n+1} - p_n)$ be negative for $n = 0, 1, 2, \cdots$. If we introduce the notation $\Delta p_n = (p_{n+1} - p_n)$ for the difference in probability of positive response in trials n and $(n + 1)$, we may write this assumption as follows:

(i) $$\Delta p_n < 0 \qquad n = 0, 1, 2, \cdots.$$

Now assumption (i) clearly gives us too much leeway. We need to know not only the *direction* of change in probabilities but also its *magnitude*. For this purpose we choose the following simple assumption: the probability of a positive response in any trial is a fixed fraction, say β, of this probability in the preceding trial. Written in symbols, we assume $p_1 = \beta p_0, p_2 = \beta p_1, \cdots$, or

(ii) $$p_{n+1} = \beta p_n \qquad n = 0, 1, 2, \cdots.$$

Since the probabilities must remain between 0 and 1, we require that

(iii) $$0 < \beta < 1.$$

Assumption (ii) may be written in the form

$$p_{n+1} - p_n = \beta p_n - p_n = (1 - \beta)(-p_n)$$

or

(iv) $$\Delta p_n = (1 - \beta)(-p_n).$$

Formula (iv) may be translated as follows: the actual decrease in probability in going from any trial to the next, Δp_n, is a fixed fraction, $(1 - \beta)$, of the maximum possible decrease, $-p_n$. (Note: if the probability is p_n and cannot fall below 0, then the maximum possible decrease is $0 - p_n$ or $-p_n$.) Assumptions (ii) and (iv) are different ways of saying the same thing; either expresses the precise way in which the probabilities p_n are postulated to vary from trial to trial. The relation in (ii) or equivalently, in (iv), is a *difference equation*. We shall use the form in (ii) for our analysis.

If we have a subject with initial probability p_0, then (ii) tells us, with $n = 0$, that

$$(0.1) \qquad\qquad p_1 = \beta p_0.$$

Now we may use (ii) again, with $n = 1$, to obtain $p_2 = \beta p_1$. With the aid of (0.1) this becomes

$$(0.2) \qquad\qquad p_2 = \beta^2 p_0.$$

Similarly, we may put $n = 2$ in (ii) and then use (0.2) to obtain

$$(0.3) \qquad\qquad p_3 = \beta p_2 = \beta^3 p_0.$$

Inspection of (0.1) through (0.3) leads us to conjecture that

$$(0.4) \qquad\qquad p_n = \beta^n p_0 \qquad n = 0, 1, 2, \cdots,$$

and indeed our later work will actually prove that (0.4) is the unique solution of the difference equation (ii) with p_0 prescribed. For the present, we take this fact as reasonable, even if not rigorously proved.

Equation (0.4) relates the subject's probability of making a positive response to three variables: n, the number of times the subject has found himself in this stimulus situation; p_0, the probability of a positive response before the experiment begins; and β, the measure of the extent to which a positive response in any trial lowers the probability of a similar response in the following trial. For example, if $p_0 = 1$ and $\beta = 0.6$, then

$$(0.5) \qquad\qquad p_n = (0.6)^n \qquad n = 0, 1, 2, \cdots,$$

so that $p_1 = 0.6$, $p_2 = 0.36$, $p_3 = 0.216$, $\cdots$. These values are graphed in Figure 0.1. Although the mathematical model gives us the values p_n for $n = 0, 1, 2, \cdots$ only, we may aid our visual comprehension of the results by drawing the indicated smooth curve through the plotted points. This is a curve of *experimental extinction* which diagrams the manner in which the probability of positive response approaches 0 under repeated trials in the absence of reward. Such a curve may be drawn for each pair of values of p_0 and β.

The problem of how to estimate the values of p_0 and β appropriate for a given experimental situation, as well as the problem of testing the usefulness of this particular analysis for summarizing data or for predictive purposes, are outside the scope of this work. These problems, together

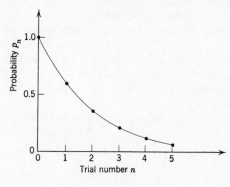

Figure 0.1

with a careful development of more elaborate probability models for learning, are treated in the recent book by R. R. Bush and F. Mosteller. References may be found in Section 2.10, where we study a more general difference equation model for learning.

Example 2

This economic example is concerned with the study of national income and its variation in time. National income, in any accounting period, is made up of three components: (1) consumer expenditures (for purchase of so-called consumer goods), (2) induced private investments (for buying capital equipment, e.g., machinery used to increase production), and (3) government expenditure. Let us introduce symbols for these quantities as follows: Y_t = national income, C_t = consumer expenditure, I_t = in-induced private investment, G_t = government expenditure. The subscript t identifies the accounting period for which the variable is evaluated. We assume that data are available for equal periods, say annually, and let t take on the values 1, 2, 3, $\cdots$, denoting the first, second, third, $\cdots$, of these periods. Our discussion so far leads to the accounting equation

$$(0.6) \qquad Y_t = C_t + I_t + G_t.$$

Following Samuelson,[1] we now make three additional assumptions relating these variables.

(i) Consumption expenditure (in any period) is proportional to the national income of the preceding period.

(ii) Induced private investment in any period is proportional to the increase in consumption of that period over the preceding (the so-called *acceleration principle*).

(iii) Government expenditure is the same in all periods.

Our problem is to analyze the behavior of national income subject to these conditions. We proceed by first restating the assumptions mathematically and then attempting to derive a single equation in which all variables but the national income are eliminated. This attempt will succeed and the resulting difference equation will enable us to study the national income as a function of time. With this as our program, we now proceed with the details.

The (positive) constant of proportionality in (i), denoted by α, is called the *marginal propensity to consume*. We therefore translate (i) by the equation

$$(0.7) \qquad\qquad C_t = \alpha Y_{t-1}.$$

Note that we have a one-period lag between income and consumption as indicated by the subscripts in (0.7).

The (positive) constant of proportionality in (ii), denoted by β, is called the *relation*. Since the increase in consumption is given by the difference $(C_t - C_{t-1})$, we rewrite (ii) in the form

$$(0.8) \qquad\qquad I_t = \beta(C_t - C_{t-1}).$$

[If consumption is decreasing, then $(C_t - C_{t-1}) < 0$ and therefore $I_t < 0$. This may be interpreted as a withdrawal from funds committed for investment purposes, for example, by not replacing depreciated machinery.]

Finally, since we are assuming in (iii) that government expenditure is the same in each period, we may as well choose our units so that this expenditure is equal to 1. Then

$$(0.9) \qquad\qquad G_t = 1$$

and equations (0.6) through (0.9) embody all our assumptions.

---

[1] P. A. Samuelson, "Interactions Between the Multiplier Analysis and the Principle of Acceleration," *Review of Economic Statistics, 21* (1939), 75–78; reprinted in *Readings in Business Cycle Theory*, Blakiston Co., Philadelphia, 1944.

To derive a single equation for national income, we start with (0.6) and use the other relations to obtain

$$Y_t = \alpha Y_{t-1} + \beta(C_t - C_{t-1}) + 1$$
$$= \alpha Y_{t-1} + \beta(\alpha Y_{t-1} - \alpha Y_{t-2}) + 1.$$

Algebraic simplification produces the *difference equation* for national income:

(0.10) $$Y_t = \alpha(1 + \beta)Y_{t-1} - \alpha\beta Y_{t-2} + 1.$$

This difference equation, relating the national income in any period to the national income of the two preceding periods, contains two parameters: the marginal propensity to consume, α, and the relation, β. Assume now that two initial values of the national income are given, say $Y_1 = 2$ and $Y_2 = 3$, and consider some special cases.

If $\alpha = 0.5$ and $\beta = 1$, then (0.10) becomes

(0.11) $$Y_t = Y_{t-1} - 0.5Y_{t-2} + 1.$$

Putting $t = 3$ and using our initial values, we obtain

$$Y_3 = Y_2 - 0.5Y_1 + 1 = 3 - (0.5)(2) + 1 = 3.$$

Now put $t = 4, 5, 6, \cdots$ in (0.11) and calculate $Y_4 = 2.5$, $Y_5 = 2.0$, $Y_6 = 1.75, \cdots$. These results are summarized in Table 0.1.

TABLE 0.1

VALUES OF Y_t, ACCORDING TO (0.10), WITH $Y_1 = 2$, $Y_2 = 3$

t	Y_t $\alpha = 0.5, \beta = 1$	Y_t $\alpha = 0.8, \beta = 2$
1	2.00	2.00
2	3.00	3.00
3	3.00	5.00
4	2.50	8.20
5	2.00	12.68
6	1.75	18.31
7	1.75	24.66
8	1.87	30.89
9	1.99	35.68
10	2.05	37.21
11	2.05	33.22
12	2.03	21.19

If we consider the case $\alpha = 0.8$ and $\beta = 2$, then (0.10) becomes

$$Y_t = 2.4\,Y_{t-1} - 1.6\,Y_{t-2} + 1$$

and with the same initial values of Y_1 and Y_2, we now compute $Y_3 = 5$, $Y_4 = 8.2, \cdots$ The first 12 values of Y_t are given in the last column of Table 0.1.

It seems that in both cases we have oscillatory behavior of the national income, the oscillations being greater in the second of the two numerical examples. But surely we have not convinced ourselves of this fact from a sequence of 12 values in each case. Is it not possible that in one of these cases, or in both, the national income stops oscillating at some time and begins a steady decrease or increase? It is clear that no table, however far extended in time, will be able to *prove* that the behavior already observed will continue. And, granting for the moment that we are certain of the behavior of Y_t when $\alpha = 0.5$ and $\beta = 1$, how does national income vary with time when $\alpha = 0.4$ and $\beta = 1$? And in what way does this behavior depend on the initial values prescribed for Y_1 and Y_2? To answer these questions, we must return to the difference equation (0.10) and use mathematical techniques which will enable us not only to identify the various possible behaviors of national income as time goes on but also to determine the values of α and β for which each of these behaviors occurs. This analysis will, for example, enable us to *prove* that if $\alpha = 0.5$ and $\beta = 1$, then if Y_1 and Y_2 are not both equal to 2, Y_t undergoes damped oscillatory motion which approaches the income value 2 as time increases, whereas for $\alpha = 0.8$ and $\beta = 2$, the income undergoes greater and greater oscillations as t increases. Nonoscillatory income behaviors are also possible for suitable values of α and β. We shall return to these considerations in Chapter 3, after having developed the necessary mathematical techniques.

In this second example, as in the first, our only concern has been to point out two facts: difference equations arise in the mathematical description of social science problems and mathematical techniques are required in order to answer relevant questions concerning the variables under analysis. The following chapters develop some elementary parts of the theory of difference equations. Wherever possible, illustrative material and exercises are selected to illustrate the applications of this theory to problems in the social sciences.

1

The
Calculus
of
Finite
Differences

1.1 THE FIRST DIFFERENCE FUNCTION

Before studying difference equations and their solutions, it is necessary to introduce the basic operations of the calculus of finite differences and to explore some properties of the corresponding difference operators. These tasks, in turn, require some background concerning the fundamental mathematical ideas of *function* and *graph*.

The study of relations among variables is a fundamental concern of all sciences. One particular kind of relation, the functional relation, is singled out here because of its major importance. Let us first consider an example or two. Given a positive real number x, denote by $A(x)$ the area of the square having a length of side equal to x. Then the equation $A(x) = x^2$ is a means of stating a rule by which we associate with each positive number x (representing length of side) another positive number $A(x)$ (representing the corresponding area). The rule is very simple in this case: to find $A(x)$, merely square the number x. That is,

$$A(2) = 2^2 = 4, \qquad A(1.3) = (1.3)^2 = 1.69, \text{ etc.}$$

Tables and graphs are other common means of indicating functional relationships. For example, the first two columns of Table 0.1 tell us the number (representing national income) to assign to each of the first 12 positive integers (the t-values).

The definition which follows makes precise our understanding of the common properties of these examples:

DEFINITION 1.1. *A function is specified when we are given a set of numbers (the domain of the function) together with a rule by which one and only one number is associated with each number in the domain. The value of a function for (or at) the number x is that number which the rule assigns to x. If the function is denoted by y, the value of y at x is denoted by $y(x)$ or y_x.*

A function may usefully be thought of as a machine for which two characteristics are known: (1) the collection of x-values that can be used as inputs of the machine, and (2) the operating rule by which the machine converts any input value to its corresponding output value.

It is important to keep clearly in mind the distinction between the function itself, denoted by y, and the numerical value, denoted by $y(x)$ or y_x, which the function rule assigns to a number x.

In our mathematical discourse we shall usually use x to represent a number in the domain of a function. However, when these numbers are to be interpreted in some particular way, we choose more natural notation. So, for example, we use t to represent a number in the domain when it is to be interpreted as a value of time, C in an economic example in which the numbers in the domain represent consumer expenditures, etc.

The functions we shall deal with are all fairly simple. In particular, the domain of our functions will usually be clear from the context and we shall therefore be able to specify a function by stating only the rule which assigns exactly one number to each element in the domain. Almost always, this rule will be specified by an equation. We agree, unless explicitly stated otherwise, to consider the domain as the largest set of numbers for which the rule is applicable, remembering that we are limited to the aggregate of real numbers.[1]

Examples

(a) Let y be the function which assigns to each real number x the constant 2, i.e., $y(x) = 2$. Such a function, whose value is the same for all numbers in its domain, is called a *constant function*.

(b) Let p be the function which assigns to each nonnegative integer n the number $(0.6)^n$. Then with p_n denoting the value of p at n,

$$p_n = (0.6)^n \qquad n = 0, 1, 2, \cdots$$

which we interpreted (p. 4) as the probability of a positive response in the nth of a series of experimental learning trials.

(c) Let y be the function specified by the equation

$$(1.1) \qquad\qquad y(x) = kx$$

[1] See E. G. Begle, *Introductory Calculus*, Henry Holt, New York, 1954, p. 44.

where k is some arbitrary number (positive, negative, or zero). The domain of y is understood to be the set of all real numbers unless some other domain is indicated. The value of y when $x = 1$ is $y(1) = k$. Similarly, $y(2) = 2k$, $y(3) = 3k, \cdots$. No matter what the value of k, any two nonzero values of y have the same ratio as the corresponding x-values. The proof of this statement is given by the equations

$$\frac{y(x_1)}{y(x_2)} = \frac{kx_1}{kx_2} = \frac{x_1}{x_2}.$$

Because of this property, y is said to be *proportional* to x and k is called the *constant of proportionality*.

If we know that y is proportional to x, this means that for *all* x-values in the domain of the function y, there is *some* number k for which (1.1) holds. Additional information, such as a known pair of values, x and $y(x)$, is required to determine the particular value of the constant k. For example, let $y(x)$ denote the time required to read x pages in this book. To say that y is proportional to x (or, in words, that total reading time is proportional to pages read) means that (1.1) holds. (Here x might sensibly be restricted to the integers $0, 1, 2, \cdots$.) If we are now told that it takes 9 minutes to read 1 page, then $y(1) = k \cdot 1 = 9$, so $k = 9$ and $y(x) = 9x$. It follows that 2 pages require 18 minutes reading time, etc.

Often a slight generalization of these ideas is needed. If y_1 and y_2 are two functions defined for the same set of x-values, and if there is some number k (the constant of proportionality) such that

$$y_1(x) = ky_2(x)$$

for all x-values in the common domain of these functions, then we shall say that y_1 *is proportional to* y_2. For example, if $C(x)$ denotes consumption expenditure and $Y(x)$ denotes income in month x (here x could be restricted to the values $1, 2, 3, \cdots$), then the assumption that consumption is proportional to income would be translated by the equation $C(x) = kY(x)$ for some constant k.

It is often helpful to have a pictorial representation or graph of a function. The framework for such a picture is a coordinate system, established when one chooses a horizontal x-axis, a vertical y-axis, and an arbitrary distance to serve as a unit of measurement along each axis. Any point on the x-axis to the right of the origin (the point of intersection of the axes) is assigned the positive number of units in the interval from the origin to the point. Points on the x-axis to the left of the origin are assigned the corresponding negative numbers. Similarly, one associates real numbers with the points on the y-axis, agreeing to assign positive numbers to points above the origin and negative numbers to those below. The origin itself is assigned the x-value 0 and the y-value 0. It is possible to show that (on each axis) each point is given one and only one real number and each real number is assigned to one and only one point.

Now if P is any point in the plane, we proceed as follows to identify it with *two* numbers. Let a vertical line be drawn through P. The intersection of this line with the x-axis determines a point whose x-value is defined to be the x-coordinate (or abscissa) of P.   Similarly, a horizontal line through P determines a point on the y-axis whose y-value is the y-coordinate (or ordinate) of P.   By this procedure, each point in the plane is assigned a pair of real numbers (x, y), and, conversely, each pair of real numbers corresponds to one and only one point.   The coordinates of a point are always written in the form (x, y) with the x-coordinate first and then the y-coordinate.   Thus $(3, 2)$ denotes the point reached by moving (from the origin) 3 units to the right on the x-axis and then 2 units up in the vertical direction; moving (again from the origin) 1 unit to the left and 5 units down yields the point $(-1, -5)$.   The origin is assigned the coordinates $(0, 0)$.   In this way, a one-to-one correspondence is established between the set of points in the plane and the set of ordered pairs of real numbers.   This correspondence is the basic idea of plane analytic geometry.

With these preliminaries understood, we can now make the following definition: *the graph of a function consists of all those points in the plane, and only those points, whose y-coordinate is the value of the function at the x-coordinate; i.e., the graph of the function y is the totality of points whose coordinates are of the form* (x, y_x), *where x is a number in the domain of y.*

Examples

(a) For the constant function given by the equation $y(x) = 2$, we know that points of the form $(x, 2)$ constitute the graph of the function.   Thus, no matter what the x-value, we must go up 2 units and the point $(x, 2)$ is on the graph. This graph is a horizontal line 2 units above the x-axis.   The graph of any constant function, say $y(x) = k$, is a horizontal line above (if $k > 0$) or below (if $k < 0$) the x-axis.   The x-axis itself is the graph of the function whose value is zero for every x.

(b) For the function given by $y(x) = x + 1$, we may plot the points $(0, 1)$, $(1, 2), (2, 3), (-1, 0)$, all of which lie on the graph since the y-value in each case satisfies the defining equation.   The graph of y is the straight line which meets the y-axis 1 unit above the x-axis, i.e., at the point $(0, 1)$, and rises 1 unit for each unit increase in x.   (The reader should note that no finite set of points, however many are plotted, can prove this last statement.   The proof is given in any analytic geometry textbook.)

(c) If $y(x) = kx$, then the graph of the function y is a straight line passing through the origin and rising (if k is positive) or falling (if k is negative) k units per unit increase in x.   If $k = 2$, for example, we may plot the points $(0, 0)$, $(1, 2), (2, 4), (-1, -2)$, etc.   For $k = -3$, the points $(0, 0), (1, -3), (2, -6)$, and $(-1, 3)$ are all on the graph of y.

(d) If y is a function whose value for any real number x is given by

$$(1.2) \qquad\qquad y_x = mx + b$$

where m and b are arbitrary constants, then y is said to be a *linear function* of x. This terminology is due to the fact that the graph of a linear function of x is a straight line. Since $y_0 = b$, this line intersects the y-axis at the point $(0, b)$, b units above (or below if b is negative) the x-axis. For this reason the number b is called the *y-intercept* of the straight line. Also

$$y_{x+1} = m(x + 1) + b = mx + m + b = y_x + m$$

so that a unit increase in x always produces a change of m units in the corresponding y-value. If m is positive, the straight line rises as x increases; if m is negative it falls as x increases; if $m = 0$, the line is horizontal. The number m is the *slope* of the straight line. A line is completely specified by giving both its y-intercept and slope. To say that y is a linear function of x is to say that there is *some* y-intercept b and *some* slope m for which (1.2) holds. Note the special cases (*a*) $m = 0, b = 2$, (*b*) $m = 1, b = 1$, (*c*) $m = k, b = 0$, already considered.

(*e*) We have already graphed the function p given by $p_n = (0.6)^n$ for $n = 0, 1, 2, \cdots$. The graph (Figure 0.1) actually consists of only the discrete points obtained for $n = 0, 1, 2, \cdots$ but we drew a smooth curve through these points, as if the domain of the function was extended to the set of all real non-negative values of n.

Very often we are given some function and are interested in studying changes in the value of the function as we move from one number to another in its domain. For example, in the learning experiment of Chapter 0, where p_n denotes the probability of a positive response in trial number n, we are interested in some measure of the amount of learning that takes place in each trial run. We may define a new function whose value at n is given by the difference between the probability of a positive response in trial $n + 1$ and trial n. Since this new function is related to p, we call it Δp so that (with Δp_n as the value of Δp at n)

$$\Delta p_n = p_{n+1} - p_n.$$

The number Δp_1 is the change in probability of positive response in going from trial 1 to trial 2, Δp_2 is the corresponding change in going from trial 2 to trial 3, etc.

As another example, if y denotes a population function, with $y(x)$ being the size of the population in census year x, we may define a new function whose value at x is the difference between the population size in years x and $x + 10$. Since this new function is related to y, we call it Δy, so that [with $\Delta y(x)$ as the value of Δy at x]

$$\Delta y(x) = y(x + 10) - y(x).$$

$\Delta y(x)$ is the change in population size in the two consecutive census years x and $x + 10$. If the population has increased, then $\Delta y(x) > 0$; if the population has declined, then $\Delta y(x) < 0$; if the population size is unchanged, then $\Delta y(x) = 0$.

Similarly, if C is the function whose value at t is denoted by C_t, then we may define another function ΔC by specifying that its value at t is the difference $(C_{t+2} - C_t)$. If the value of ΔC at t is denoted by ΔC_t, then

$$\Delta C_t = C_{t+2} - C_t$$

and we may choose to interpret this as the difference between consumer expenditures in the 2-year interval between periods t and $t + 2$.

In the first of these examples our difference interval was 1, in the second 10, in the third 2. This interval may be any number provided both the

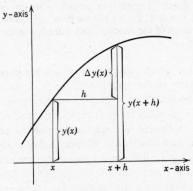

Figure 1.1

initial and final values at which the function is to be evaluated are in its domain. These considerations are formalized in the following definition:

DEFINITION 1.2. *Let a function y be given and let h be any constant for which x + h is in the domain of y whenever x is. Then Δy, the first difference of y, is that function whose value at x, denoted by $\Delta y(x)$ (or Δy_x), is given[2] by*

(1.3) $$\Delta y(x) = y(x + h) - y(x).$$

The symbol Δ denotes the *difference operator*, indicating that the function y is to be operated upon (or transformed) to yield the new function Δy. The number h is called the *difference interval*.

---

[2] The first difference is often called the *forward* difference of y to distinguish it from the *backward* difference, whose value at x is $y(x) - y(x - h)$. Unfortunately, notation is not completely uniform so that in some references, especially in economics, the symbol Δy is used for the backward difference.

We shall understand, unless it is specifically indicated otherwise, that we are always using the same difference interval h. Because of this agreement, we shall not explicitly indicate the dependence of Δy on h. (Some authors write Δy_h in order to avoid all possibility of misunderstanding.)   For the simple function given by $y(x) = x$, we find, using (1.3),

$$\Delta y(x) = \Delta x = (x + h) - x$$

or

$$\Delta x = h.$$

For this reason, the difference interval is often denoted by Δx.

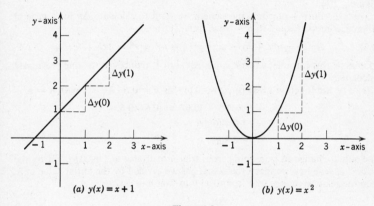

(a) $y(x) = x + 1$ (b) $y(x) = x^2$

Figure 1.2

A graphical analysis may be helpful here.   In Figure 1.1, as one moves from x to $x + h$ in the domain of y, the value of y changes from $y(x)$ to $y(x + h)$.   The first difference of y may be interpreted as the function whose value at x is the increment or change in y which corresponds to the increment h (or Δx) in x.

Examples

(a) If y is specified by $y(x) = x + 1$, and we take the difference interval h equal to 1, then some particular values of Δy are obtained as follows:

$$\Delta y(1) = y(2) - y(1) = 3 - 2 = 1,$$
$$\Delta y(2.5) = y(3.5) - y(2.5) = 4.5 - 3.5 = 1, \text{ etc.}$$

As a matter of fact, we may prove that the value of Δy is 1 for every value of x. For

$$\Delta y(x) = y(x + 1) - y(x) = (x + 2) - (x + 1) = 1.$$

This result is not surprising since the graph of the function y is a straight line for which, independently of the starting value x, each unit change in x produces the same change in the value of y. See Figure 1.2(a).

(b) If $y(x) = x^2$, then (again taking $h = 1$)

$$\Delta y(1) = y(2) - y(1) = 2^2 - 1^2 = 3,$$
$$\Delta y(2.5) = y(3.5) - y(2.5) = (3.5)^2 - (2.5)^2 = 6, \text{ etc.}$$

In general,

$$\Delta y(x) = y(x + 1) - y(x) = (\ + 1)^2 - x^2$$
$$= (x^2 + 2x + 1) - x^2$$
$$= 2x + 1.$$

For this function, the value of Δy is not constant, but depends upon x. The increment in y caused by moving from x to $x + 1$ increases as x increases. See Figure 1.2(b).

The procedure is precisely the same if we want to calculate Δy for a general difference interval equal to h. Now, using (1.3),

(1.4) $$\Delta y(x) = y(x + h) - y(x) = (x + h)^2 - x^2 = 2xh + h^2.$$

Of course, this formula, when h is put equal to 1, reduces to the one previously calculated.

(c) For the function given by $p_n = (0.6)^n$ for $n = 0, 1, 2, \cdots$, with $h = 1$,

$$\Delta p_n = (0.6)^{n+1} - (0.6)^n = (0.6)^n(0.6 - 1)$$
$$= (-0.4)(0.6)^n$$
$$= (-0.4)p_n.$$

The actual change in p in going from trial n to trial $n + 1$ is Δp_n. If by the *relative change* in p we mean the actual change divided by the initial value of p, then the relative change in p from trial n to trial $n + 1$ is

$$\frac{\Delta p_n}{p_n} = -0.4$$

i.e., this relative change is a constant from trial to trial. It is negative or a *decrease* of numerical value 0.4 or 40%. [Cf. assumptions (i) and (iv) in Example 1 of Chapter 0.]

PROBLEMS 1.1

In Problems 1–10, a function y is specified by giving its value $y(x)$ for any number x. Find (a) $y(0)$, $y(1)$, $y(2)$; (b) $\Delta y(0)$, $\Delta y(1)$, $\Delta y(2)$, $\Delta y(x)$, assuming $h = 1$; (c) $\Delta y(x)$ with a difference interval equal to h. The symbols a, b, and c denote constants.

1. $y(x) = 1$
2. $y(x) = x - 2$
3. $y(x) = 3x + 5$
4. $y(x) = ax + b$
5. $y(x) = x^2 + 1$
6. $y(x) = x(x - h)$
7. $y(x) = ax^2 + bx + c$
8. $y(x) = x(x - h)(x - 2h)$
9. $y(x) = 2^x$
10. $y(x) = 3^x$

11. Sketch the graphs of the functions in Problems 1–5.

12. Consider the functions given by $y_1(x) = x$ and $y_2(x) = x^2$. Show that $\Delta y_1(x) = \Delta x = h$ and $\Delta y_2(x) = \Delta x^2 = 2xh + h^2$. Note the difference in meaning (and value) between Δx^2 and $(\Delta x)^2$.

13. Consider the following table:[3]

x	0	1	2	3	4	5	6	7
$y(x)$	A	A	A	B	B	C	C	C

where A, B, and C are numbers. Verify directly from the definition that y is a function of x. What is the domain of definition of this function?

14. Let $I(t)$ be the intensity of light required to produce a certain fixed photochemical decomposition on an illuminated substance in time t. Bartley[4] derives the equation

$$I(t) = C \cdot \frac{1}{t} + D \qquad t > 0$$

where C and D are constants. Put $C = 1$ and $D = 0$ and make a rough sketch of the graph of the function I. As t increases, note that $I(t)$ decreases. Conclude that $\Delta I(t)$ is negative. Finally, show that your conclusion is not changed if D is different from 0.

15. Let the function y be defined with domain the set of x-values between 0 and 1 inclusive, and suppose that

$$y(x) = x(1 - x).$$

Show algebraically that

$$y(x) = \tfrac{1}{4} - (x - \tfrac{1}{2})^2$$

and thus conclude that the largest value of y is $\tfrac{1}{4}$ and that this value is assumed when $x = \tfrac{1}{2}$.

16. In a certain population, let y_t denote the *proportion* of the articulate population who openly favor war at time t. The proportion who do not openly favor war at time t is therefore $1 - y_t$. Richardson[5] assumes that Δy_t is proportional to the product $y_t(1 - y_t)$, i.e.,

$$\Delta y_t = ky_t(1 - y_t)$$

for some constant k. Note that y_t, being a proportion, is a number between 0 and 1 inclusive, and use the result of the preceding problem to show that the change in proportion of those who favor war at time $t + \Delta t$ as compared with time t is greatest when $y_t = \tfrac{1}{2}$, i.e., when half the population openly favors war.

[3] See S. A. Stouffer et al., *Measurement and Prediction*, Vol. IV of *Studies in Social Psychology in World War II*, Princeton Univ. Press, Princeton, 1950, pp. 63–65.

[4] S. H. Bartley, "The Psychophysiology of Vision," Chap. 24 in S. S. Stevens (ed.), *Handbook of Experimental Psychology*, John Wiley, New York, 1951. Cf. equation 6, p. 940.

[5] L. F. Richardson, "War-moods: I," *Psychometrika*, *13* (1948), 147–174. The differential calculus is used in this paper.

17. The demand function in economics specifies the quantity of a certain commodity that people will buy at each price at which it is offered. If p denotes price, we let $D(p)$ denote the quantity demanded at price p. Express the following assumption in suitable mathematical notation: an increase in price causes a fall in the quantity demanded. If the demand function is linear, what does the assumption tell us of the slope of the straight-line graph of this function?

18. The supply function specifies the quantity of a certain commodity that will be offered for purchase at a given price. Let $S(p)$ denote this supply at price p. Express in suitable mathematical notation the assumption that an increase in price causes an increase in the quantity supplied. If the supply function is linear, show that the slope of its straight-line graph is positive.

19. The supply and demand functions for sugar from 1890 to 1915 were estimated[6] to be given by

$$S(p) = 0.7p + 0.4 \qquad D(p) = 1.6 - 0.5p.$$

If the market price is defined as that value of p at which supply and demand are equal, show that the market price is $p = 1$. What quantity is demanded at this market price? Draw the graphs of the supply and demand functions on the same set of axes and interpret the point of intersection of the resulting straight lines.

20. If $D(p)$ is the quantity demanded at the price p, then $D(p + \Delta p)$ is demanded at the price $p + \Delta p$ and $\Delta D(p)$ denotes the difference in demand due to this price change of amount Δp. The quantity

$$\frac{p}{D(p)} \bigg/ \frac{\Delta p}{\Delta D(p)} = \frac{p}{D(p)} \cdot \frac{\Delta D(p)}{\Delta p}$$

is referred to as the *average* or *arc elasticity*[7] of demand for the price interval p to $p + \Delta p$. With the assumption of Problem 17, show that this average elasticity is a negative number. If the demand function is linear, say $D(p) = mp + b$, show that $\Delta D(p) = m\Delta p$ and conclude that the average elasticity, although dependent upon the initial price p, does *not* depend on the price change, Δp.

1.2 SECOND AND HIGHER DIFFERENCES

Starting with a function y, we defined the new function Δy which results when the operator Δ is applied to (the operand) y. We can, if we like, apply the operator Δ to this new function Δy. Let us consider a simple example. If y is specified by the equation $y(x) = x^2$, then we find, in (1.4), that Δy is given by $\Delta y(x) = 2xh + h^2$, where h is the interval of

---

[6] H. Schultz, *Statistical Laws of Demand and Supply with Special Applications to Sugar*, Univ. of Chicago Press, Chicago, 1928.

[7] R. G. D. Allen, "The Concept of Arc Elasticity of Demand," *Review of Economic Studies*, 1 (1934), 226–229.

differencing. Now $\Delta(\Delta y)$ is that function, according to Definition 1.2, whose value at x is $[\Delta y(x + h) - \Delta y(x)]$. But

$$\Delta y(x) = 2xh + h^2,$$

so that

$$\Delta y(x + h) = 2(x + h)h + h^2 = 2xh + 3h^2$$

and

$$\Delta y(x + h) - \Delta y(x) = 2h^2.$$

Thus the value of the function $\Delta(\Delta y)$ for any x is the constant $2h^2$. The notation $\Delta^2 y$ is used to denote this function and its value at x is denoted by $\Delta^2 y(x)$. We have found

$$\Delta^2 y(x) = \Delta^2 x^2 = 2h^2.$$

Undaunted, we march ahead and obtain yet another function by applying the operator Δ to the function $\Delta^2 y$. This function we call $\Delta^3 y$ and, again from Definition 1.2, we know that the value of $\Delta^3 y$ at x is $[\Delta^2 y(x + h) - \Delta^2 y(x)]$. Having found that $\Delta^2 y(x)$ is constant $(= 2h^2)$, we know

$$\Delta^3 y(x) = \Delta^2 y(x + h) - \Delta^2 y(x) = 2h^2 - 2h^2 = 0.$$

We may continue this process indefinitely. Since $\Delta^3 y$ is zero for every x, the function $\Delta^4 y$ will also be identically zero. In fact, $\Delta^5 y$, $\Delta^6 y$, $\cdots$ will all be constant functions $(= 0$ for every $x)$ when the function y is given by $y(x) = x^2$.

DEFINITION 1.3. *Suppose a function y and its first difference Δy are given. Then the second difference of y, denoted by $\Delta^2 y$, is the difference of the first difference of y, i.e.,*

$$(1.5) \qquad \Delta^2 y = \Delta(\Delta y).$$

[The value of the function $\Delta^2 y$ at x, denoted by $\Delta^2 y(x)$, is therefore given by

$$(1.6) \qquad \Delta^2 y(x) = \Delta y(x + h) - \Delta y(x).]$$

Similarly, the third difference of y, denoted by $\Delta^3 y$, is the difference of the second difference of y, i.e.,

$$\Delta^3 y = \Delta(\Delta^2 y),$$

and, in general, the nth difference of y, denoted by $\Delta^n y$, is the difference of the $(n - 1)$st difference of y, i.e.,

$$(1.7) \qquad \Delta^n y = \Delta(\Delta^{n-1} y) \qquad n = 2, 3, 4, \cdots.$$

Note that (1.7), when $n = 2$, yields the identity $\Delta^2 y = \Delta(\Delta^1 y)$. In order that this be consistent with (1.5), we agree to omit the superscript when writing the first difference of a function; i.e., we write Δy (as we have been doing all along) rather than $\Delta^1 y$. One other convention is useful since it enables (1.7) to be meaningful for $n = 1$. With $n = 1$, (1.7) reads $\Delta y = \Delta(\Delta^0 y)$. If this is to be correct, we must agree that $\Delta^0 y = y$; i.e., Δ^0 is the operator which, when applied to a function y, leaves y unchanged. Such an operator is called an identity operator.

DEFINITION 1.4. *The identity operator, denoted by I, is that operator which, applied to any function y, produces a new function Iy identical with y. That is, for any x in the domain of y, we have*

$$(1.8) \qquad\qquad Iy(x) = y(x).$$

The symbol Δ^0 is defined as the identity operator, i.e.,

$$(1.9) \qquad\qquad \Delta^0 y = Iy$$

and, as was our intent in making this particular definition, (1.7) is now correct for $n = 1$.

Example

Let y be the function for which $y(x) = x^3$. Then

$$\Delta y(x) = \Delta x^3 = (x + h)^3 - x^3.$$

We apply the binomial theorem[8] to expand $(x + h)^3$ as follows:

$$(x + h)^3 = x^3 + 3x^2 h + 3xh^2 + h^3.$$

Then

$$\Delta y(x) = 3x^2 h + 3xh^2 + h^3.$$

In comparing the function values $y(x)$ and $\Delta y(x)$, we note that the highest power of x has been reduced by 1. The function y is cubic in x (highest power of x appearing is 3) and Δy is a quadratic function of x (highest power of x is 2). Now

$$\begin{aligned}
\Delta^2 y(x) &= \Delta y(x + h) - \Delta y(x) \\
&= [3(x + h)^2 h + 3(x + h)h^2 + h^3] - (3x^2 h + 3xh^2 + h^3) \\
&= (3x^2 h + 9xh^2 + 7h^3) - (3x^2 h + 3xh^2 + h^3) \\
&= 6xh^2 + 6h^3.
\end{aligned}$$

Again we have lowered the degree of our function by applying the operator Δ, for $\Delta^2 y$ is a linear function of x (highest power of x is 1). Finally,

$$\begin{aligned}
\Delta^3 y(x) &= \Delta^2 y(x + h) - \Delta^2 y(x) \\
&= [6(x + h)h^2 + 6h^3] - (6xh^2 + 6h^3) \\
&= 6h^3, \text{ a constant,}
\end{aligned}$$

[8] For any positive integer n,

$$(A + B)^n = A^n + \frac{n}{1} A^{n-1} B + \frac{n(n-1)}{1 \cdot 2} A^{n-2} B^2 + \frac{n(n-1)(n-2)}{1 \cdot 2 \cdot 3} A^{n-3} B^3 + \cdots + B^n.$$

The sum on the right has $(n + 1)$ terms and is said to be the *expansion* of $(A + B)^n$.

so that

$$\Delta^4 y(x) = \Delta^3 y(x + h) - \Delta^3 y(x)$$
$$= 6h^3 - 6h^3 = 0,$$

and, of course, all higher differences are likewise zero.

We may summarize these results, as well as some others, in Table 1.1.

TABLE 1.1

$y(x)$	$\Delta y(x)$	$\Delta^2 y(x)$	$\Delta^3 y(x)$	$\Delta^n y(x)$, $n \geq 4$
1	0	0	0	0
x	h	0	0	0
x^2	$2xh + h^2$	$2h^2$	0	0
x^3	$3x^2 h + 3xh^2 + h^3$	$6xh^2 + 6h^3$	$6h^3$	0

We have found a number of functions for which repeated applications of the operator Δ eventually yield a function $\Delta^n y$ whose values are constant, so that all higher differences of y are identically zero. In Section 1.4 we shall show that these results are special cases of a more general theorem.

PROBLEMS 1.2

1–10. For each of the functions in 1–10 of Problems 1.1, find (a) $\Delta^2 y(1)$ and $\Delta^2 y(x)$, assuming $h = 1$; (b) $\Delta^2 y(x)$ with a difference interval equal to h.

11. If $y(x) = 2^x$ and $c = 2^h - 1$, show that $\Delta y(x) = cy(x)$, $\Delta^2 y(x) = c^2 y(x)$, $\Delta^3 y(x) = c^3 y(x)$. What is your conjecture for $\Delta^n y(x)$ for any positive integer n?

12. Show that

$$\Delta^2 y(x) = y(x + 2h) - 2y(x + h) + y(x),$$
$$\Delta^3 y(x) = y(x + 3h) - 3y(x + 2h) + 3y(x + h) - y(x).$$

1.3 THE OPERATOR E

The operator Δ applied to a function y requires that we perform two operations to get $\Delta y(x)$: we change $y(x)$ to $y(x + h)$ and then subtract $y(x)$ from the result. It turns out to be extremely useful to introduce a special symbol for the first of the operations.

DEFINITION 1.5. *Let a function y be given and let h be any constant. Then Ey is that function whose value at x, denoted by $Ey(x)$ (or Ey_x), is given by*

(1.10) $$Ey(x) = y(x + h).$$

Of course, the number $(x + h)$ must be in the domain of y. We can now write

(1.11) $$\Delta y(x) = Ey(x) - y(x).$$

The operator E, like Δ, may be applied more than once to a function. So, for example, with notation similar to that employed in Definition 1.3,

$$E^2y(x) = E[Ey(x)] = Ey(x + h) = y(x + 2h),$$
$$E^3y(x) = E[E^2(x)] = Ey(x + 2h) = y(x + 3h),$$

and, in general, we can show that with the definition

(1.12) $$E^ny(x) = E[E^{n-1}y(x)] \qquad n = 1, 2, 3, \cdots$$

we have

(1.13) $$E^ny(x) = y(x + nh) \qquad n = 1, 2, 3, \cdots.$$

Before proving this, let us agree to write E rather than E^1. As with the operator Δ, in order that (1.12) be correct for $n = 1$, we must make the definition:

(1.14) $$E^0y(x) = Iy(x) = y(x).$$

With these conventions, (1.13) is correct for $n = 0, 1, 2, \cdots$.

Examples

(a) If $y(x) = 3x$, then $Ey(x) = y(x + h) = 3(x + h) = 3x + 3h$. From (1.13), $E^ny(x) = y(x + nh) = 3(x + nh) = 3x + 3nh$.

(b) If $y(x) = 2^x$, then $E^0y(x) = 2^x$, $Ey(x) = y(x + h) = 2^{x+h} = 2^h2^x$, and in general $E^ny(x) = y(x + nh) = 2^{x+nh} = 2^{nh}2^x$.

(c) Let C be a function whose value, C_t, at the positive integer t denotes consumer expenditure in period t. Then (with $h = 1$), $EC_t = C_{t+1}$ is the consumer expenditure in the following period. The equations $EC_t = C_t$ ($t = 1, 2, \cdots$) assert that consumer expenditure is constant from period to period.

To prove (1.13) is true for any positive integer, we make use of the method of proof known as *mathematical induction*. Although most readers will be familiar with this method, it is important enough for our purposes to justify the following very brief review.

Let $P(n)$ denote a proposition involving the positive integer n. If n is put equal to $1, 2, 3, \cdots$, we obtain the propositions $P(1), P(2), P(3), \cdots$. Each of these propositions is either true or false. For example, if

$$P(n):\text{ the number } n^2 \text{ is less than 5,}$$

then $P(1)$ and $P(2)$ are true statements, but $P(3), P(4), \cdots$ are all false.

To prove, by mathematical induction, that P(n) is true for all positive integers n, it suffices to prove two statements: (1) *the proposition P(1) is true, and* (2) *for any positive integer k, if P(k) is true, then P(k + 1) is also true.*

Step (1) verifies the truth of $P(n)$ when $n = 1$; step (2) shows that the truth of the proposition for any integer follows from its truth for the preceding integer. Taken together, these steps prove (by mathematical induction) that $P(n)$ is true for any positive integer. (It is as if to prove we may reach *any* rung on an infinitely extended ladder, we first show that we may step onto the first rung, and then show that we are able to step from any rung to the next higher one.)

The following examples illustrate this method of proof.

Examples

(*a*) To prove that

$$P(n): \text{ the sum of the first } n \text{ positive integers is } \frac{n(n + 1)}{2},$$

or, in symbols, that

$$P(n): 1 + 2 + 3 + \cdots + n = \frac{n(n + 1)}{2}$$

is true for all values of n, we construct the required two-step proof. (1) $P(1)$ is the statement $1 = [1(1 + 1)]/2$ and this is clearly true. (2) We assume $P(k)$ is true and must prove $P(k + 1)$ is also true. Now

$$P(k): 1 + 2 + 3 + \cdots + k = \frac{k(k + 1)}{2}$$

being true (by hypothesis), we obtain another true statement by adding $(k + 1)$ to both sides of the equation. That is,

$$1 + 2 + 3 + \cdots + k + (k + 1) = \frac{k(k + 1)}{2} + k + 1$$

$$= (k + 1)\left(\frac{k}{2} + 1\right)$$

$$= \frac{(k + 1)(k + 2)}{2}.$$

But this final statement is just $P(k + 1)$. Hence $P(k + 1)$ is a consequence of $P(k)$ and the proof is complete.

(*b*) We prove that

$$P(n): E^n y(x) = y(x + nh)$$

is true for all positive integral values of n. This is (1.13). (1) $P(1)$ is true. For

$$P(1): E^1 y(x) = y(x + h)$$

and this is true by Definition 1.5 and our agreement to identify E^1 with E. (2) We now assume

$$P(k): E^k y(x) = y(x + kh)$$

is true, and attempt to prove that $P(k + 1)$ is true. We apply the operator E to both sides of the equation $P(k)$ to get

$$E[E^k y(x)] = Ey(x + kh).$$

Now we use the definition of E^{k+1} given by (1.12), as well as the definition of the operator E, to rewrite this in the form

$$E^{k+1}y(x) = y(x + kh + h)$$
$$= y[x + (k + 1)h].$$

But the final equality is the proposition $P(k + 1)$ obtained by replacing n in $P(n)$ by $k + 1$. Hence we have proved that $P(k)$ implies $P(k + 1)$ and step (2) [and therefore the proof of (1.13) by mathematical induction] is complete.

PROBLEMS 1.3

1–4. For each of the functions in 1–4 of Problems 1.1 find (a) $Ey(x)$; (b) $E^2y(x)$; (c) $E^ny(x)$. Use a difference interval equal to h.

5. If $y(x) = 3^x$, show that $E^ny(x) = 3^{nh}y(x)$ for $n = 0, 1, 2, \cdots$.

6. Show that (cf. 12 of Problems 1.2)

(1.15)
$$\Delta^2y(x) = E^2y(x) - 2Ey(x) + Iy(x)$$
$$\Delta^3y(x) = E^3y(x) - 3E^2y(x) + 3Ey(x) - Iy(x).$$

7. Show that

(1.16)
$$E^2y(x) = y(x) + 2\Delta y(x) + \Delta^2 y(x)$$
$$E^3y(x) = y(x) + 3\Delta y(x) + 3\Delta^2 y(x) + \Delta^3 y(x).$$

8. Let p be the function whose value, p_n, at the positive integer n is interpreted as the probability of a positive response in the nth experimental trial. What is the interpretation of Ep_n? If $p_n = (0.6)^n$, show that $Ep_n = (0.6)p_n$. [Cf. assumption (ii) in Example 1 of Chapter 0.]

9. If we define the operator I^n for $n = 2, 3, 4, \cdots$ in the same recursive manner used to define Δ^n and E^n, i.e., $I^ny = I(I^{n-1})y$, then show using mathematical induction that $I^ny(x) = Iy(x)$.

1.4 SOME PROPERTIES OF Δ AND E

Since the operators Δ and E are of fundamental importance, we should like to be able to calculate the values of Δy and Ey with some degree of ease, at least for the functions y which arise fairly often. Table 1.1 gives us the results of applying Δ to four simple functions. But what if we must apply Δ to some combination of these functions, say $y(x) = x + x^2$? Now

$$\Delta y(x) = y(x + h) - y(x)$$
$$= [(x + h) + (x + h)^2] - (x + x^2)$$
$$= [(x + h) - x] + [(x + h)^2 - x^2]$$
$$= \Delta x + \Delta x^2.$$

Thus we find

$$\Delta(x + x^2) = \Delta x + \Delta x^2.$$

Had we known this fact (namely, the difference of the sum[9] of two functions is equal to the sum of the differences of the functions) we could have avoided the above algebra and obtained $\Delta(x + x^2)$ directly from Table 1.1.

Such considerations lead to the search for formulas by which, having calculated $\Delta y(x)$ directly from Definition 1.2 for a number of functions y, we may find the differences of still other functions by using the results of these calculations rather than again resorting to the basic definition. If the differences of a large class of functions can be obtained with relatively few such formulas and if these are fairly simple to use, then we certainly have a net gain. Some of these formulas of the calculus of finite differences will be developed here.

THEOREM 1.1. *The Commutative Law with Respect to Constants. If c is any constant, the difference of the function cy is equal to c times the difference of y, i.e.,*

$$(1.17) \qquad \Delta[cy(x)] = c\Delta y(x).$$

PROOF. We use Definition 1.2 to write

$$\Delta[cy(x)] = cy(x + h) - cy(x)$$

$$= c[y(x + h) - y(x)]$$

$$= c\Delta y(x).$$

Formula (1.17) enables us to use the results of Table 1.1 to find the following differences with essentially no calculation:

$$\Delta(3x) = 3\Delta x = 3h, \qquad \Delta(5x^2) = 5\Delta x^2 = 5(2xh + h^2), \text{ etc.}$$

THEOREM 1.2. *The Distributive Law. The difference of the sum of two functions is equal to the sum of their differences, i.e., if y_1 and y_2 are two functions,*

$$(1.18) \qquad \Delta[y_1(x) + y_2(x)] = \Delta y_1(x) + \Delta y_2(x).$$

---

[9] If y_1 and y_2 are two functions, then the sum of these functions is a new function whose domain consists of those numbers in *both* the domain of y_1 and y_2; if x is such a number, then the sum function has, by definition, the value $y_1(x) + y_2(x)$. Similar natural definitions are given for the difference, product, and quotient of two functions. Also, two functions are said to be equal if they have the same domain of definition and the same value for each number in this common domain. For full details and examples, see E. G. Begle, *Introductory Calculus*, Henry Holt, New York, 1954, especially pp. 42–47, 56–58.

PROOF.   From Definition 1.2,

$$\Delta[y_1(x) + y_2(x)] = [y_1(x + h) + y_2(x + h)] - [y_1(x) + y_2(x)]$$
$$= [y_1(x + h) - y_1(x)] + [y_2(x + h) - y_2(x)]$$
$$= \Delta y_1(x) + \Delta y_2(x).$$

COROLLARY 1.   *If c_1 and c_2 are arbitrary constants,*

(1.19) $$\Delta[c_1 y_1(x) + c_2 y_2(x)] = c_1 \Delta y_1(x) + c_2 \Delta y_2(x).$$

PROOF.   Apply Theorem 1.2 to the functions $c_1 y_1$ and $c_2 y_2$ and obtain

$$\Delta[c_1 y_1(x) + c_2 y_2(x)] = \Delta[c_1 y_1(x)] + \Delta[c_2 y_2(x)].$$

The result follows by applying (1.17) to the two terms on the right.

COROLLARY 2.   *Let y_1, y_2, $\cdots$, y_n be n functions and c_1, c_2, $\cdots$, c_n be arbitrary constants.   Then for any positive integer n,*

(1.20) $$\Delta[c_1 y_1(x) + c_2 y_2(x) + \cdots + c_n y_n(x)]$$
$$= c_1 \Delta y_1(x) + c_2 \Delta y_2(x) + \cdots + c_n \Delta y_n(x).$$

The result is true for $n = 1$ by (1.17) and for $n = 2$ by (1.19).   By assuming (1.20) true for $n = k$ and proving it true for $n = k + 1$, which we omit here, the proof of Corollary 2 by mathematical induction can be completed.

This brief digression concerns a convenient notation, to be used henceforth, for writing sums like those in (1.20).   If y_k denotes the value of a function y at the integer value k and if a and b are integers ($b \geq a$), the sum

$$y_a + y_{a+1} + y_{a+2} + \cdots + y_b$$

is instead written in the condensed form

$$\sum_{k=a}^{b} y_k.$$

This symbol is read "the sum of y_k from $k = a$ to $k = b$."   Some examples will help to clarify the use of this summation notation.   In each example, the sum is also written out in long form and described in words.

Examples

   (*a*) The sum of the first n positive integers:

$$\sum_{k=1}^{n} k = 1 + 2 + 3 + \cdots + n.$$

   (*b*) The sum of the first ten odd integers:

$$\sum_{k=0}^{9} (2k + 1) = 1 + 3 + 5 + \cdots + 19.$$

(c) The sum of the first, second, and third differences of the function y, each evaluated at x:

$$\sum_{k=1}^{3} \Delta^k y(x) = \Delta y(x) + \Delta^2 y(x) + \Delta^3 y(x).$$

It is convenient to know the following rules concerning summation. We let y and z be any two functions and c an arbitrary constant. Then

(1) $\displaystyle\sum_{k=1}^{n} c = nc.$

(2) $\displaystyle\sum_{k=1}^{n} c y_k = c \sum_{k=1}^{n} y_k.$

(3) $\displaystyle\sum_{k=1}^{n} (y_k \pm z_k) = \sum_{k=1}^{n} y_k \pm \sum_{k=1}^{n} z_k.$

We prove (2) and leave the others for the reader. The proof is given by

$$\sum_{k=1}^{n} c y_k = c y_1 + c y_2 + c y_3 + \cdots + c y_n = c(y_1 + y_2 + y_3 + \cdots + y_n)$$

$$= c \sum_{k=1}^{n} y_k.$$

As an example of the use of these rules, let $y_1, y_2, \cdots, y_n$ be n measurements of some variable (weights, IQ scores, etc.). The arithmetic mean of these n measurements is

$$\bar{y} = \frac{y_1 + y_2 + \cdots + y_n}{n} = \frac{1}{n} \sum_{k=1}^{n} y_k.$$

An important property of the arithmetic mean is the following: the sum of the deviations of the measurements from their arithmetic mean is 0, i.e.,

$$\sum_{k=1}^{n} (y_k - \bar{y}) = 0.$$

To prove this result, write

$$\sum_{k=1}^{n} (y_k - \bar{y}) = \sum_{k=1}^{n} y_k - \sum_{k=1}^{n} \bar{y} \qquad \text{[by rule (3)]}$$

$$= \sum_{k=1}^{n} y_k - n\bar{y} \qquad \text{[by rule (1)]}$$

$$= 0 \qquad \text{(by definition of } \bar{y}\text{)}.$$

Now we can use the summation notation[10] to rewrite (1.20) in the form

(1.20') $$\Delta \sum_{k=1}^{n} c_k y_k(x) = \sum_{k=1}^{n} c_k \Delta y_k(x)$$

---

[10] For future reference, we remark that this notation can be used to write the binomial formula as

$$(A + B)^n = \sum_{k=0}^{n} \frac{n(n-1)(n-2) \cdots (n-k+1)}{1 \cdot 2 \cdot 3 \cdots k} A^k B^{n-k}.$$

Cf. footnote 8.

Formula (1.20') enables us to find the difference of any linear combination of functions provided that we can find the differences of the functions themselves.

Examples

(a) We may use Table 1.1 and (1.19) to obtain

$$\Delta(3x^2 + 5x) = 3\Delta x^2 + 5\Delta x = 3(2xh + h^2) + 5h$$

(b) By (1.20'), with a, b, and c constants (cf. 7 of Problems 1.1)

$$\Delta(a + bx + cx^2) = a\Delta 1 + b\Delta x + c\Delta x^2$$
$$= a \cdot 0 + b \cdot h + c(2xh + h^2)$$
$$= bh + ch^2 + 2chx.$$

(c) Suppose we have a function C whose value at N, denoted by C_N, is given by

$$C_N = \frac{a + bN}{1 - b} = \frac{a}{1 - b} + \frac{b}{1 - b} N,$$

where a and b are constants, $b \neq 1$. Then

$$\Delta C_N = \Delta\left(\frac{a}{1 - b}\right) + \Delta\left(\frac{b}{1 - b} N\right) = 0 + \frac{b}{1 - b} \Delta N$$

so that[11]

$$\frac{\Delta C_N}{\Delta N} = \frac{b}{1 - b}.$$

We may use (1.20) to prove the following very important result.

THEOREM 1.3. *Let y be a polynomial of degree n, i.e.,*

(1.21) $$y(x) = a_0 + a_1 x + a_2 x^2 + \cdots + a_n x^n$$

with $a_0, a_1, \cdots, a_n$ arbitrary constants and $a_n \neq 0$. Then the nth difference of y is a constant function (equal to $n!h^n a_n$) and all succeeding differences are zero, i.e.,

(1.22) $$\Delta^p y(x) = 0 \qquad \text{if } p > n.$$

PROOF. By (1.20)

(1.23) $$\Delta y(x) = a_0\Delta 1 + a_1\Delta x + a_2\Delta x^2 + \cdots + a_n\Delta x^n.$$

---

[11] See Thomas C. Schelling, *National Income Behavior*, McGraw-Hill, New York, 1951, p. 38. There C and N are interpreted as consumer and nonconsumer expenditures respectively.

But if m is any positive integer, the binomial theorem may be used to obtain

$$\Delta x^m = (x + h)^m - x^m$$

$$= x^m + mx^{m-1}h + \frac{m(m-1)}{2} x^{m-2}h^2 + \cdots + h^m - x^m$$

$$= mx^{m-1}h + \frac{m(m-1)}{2} x^{m-2}h^2 + \cdots + h^m;$$

i.e., application of Δ to x^m produces a finite number of terms with x^{m-1} as the highest power of x appearing. So by applying Δ to $y(x)$ we get, using this fact and (1.23), a sum of terms, each a constant times a power of x and the highest power of x appearing is $(n - 1)$. In other words, *application of Δ to a polynomial of degree n results in a polynomial of degree $(n - 1)$.* By the same reasoning, $\Delta^2 y$ will be a polynomial of degree $(n - 2)$ and after n applications of Δ we are reduced to a polynomial of degree 0 or a constant. This proves that $\Delta^n y$ is a constant and the final part of the theorem follows immediately since the difference of a constant is zero. The evaluation of $\Delta^n y$ is outlined in 6–8 of Problems 1.4. Theorem 1.3 is of great importance in the theory of polynomial interpolation. (Cf. Section 1.5.)

We are severely limited in using the formulas already developed due to the fact that we have at our disposal such a small collection of functions whose differences we know. Instead of extending Table 1.1 by considering $x^4, x^5, \cdots$, we introduce another useful set of functions.

DEFINITION 1.6. *The factorial function of order n ($n = 1, 2, 3, \cdots$) is that function whose value at x is given by*

$$x^{(n)} = x(x - h)(x - 2h) \cdots [x - (n - 1)h].$$

For example,

$$x^{(1)} = x \qquad\qquad (x + h)^{(1)} = x + h$$
$$x^{(2)} = x(x - h) \qquad\qquad (x + h)^{(2)} = (x + h)x$$
$$x^{(3)} = x(x - h)(x - 2h) \qquad (x + h)^{(3)} = (x + h)(x)(x - h),$$

so that, if $h = 2$,

$$3^{(1)} = 3 \qquad 3^{(2)} = 3 \cdot 1 = 3 \qquad 3^{(3)} = 3 \cdot 1 \cdot -1 = -3$$
$$4^{(1)} = 4 \qquad 4^{(2)} = 4 \cdot 2 = 8 \qquad 4^{(3)} = 4 \cdot 2 \cdot 0 = 0,$$

whereas, if $h = 3$,

$$3^{(1)} = 3 \qquad 3^{(2)} = 3 \cdot 0 = 0 \qquad 3^{(3)} = 3 \cdot 0 \cdot -3 = 0$$
$$4^{(1)} = 4 \qquad 4^{(2)} = 4 \cdot 1 = 4 \qquad 4^{(3)} = 4 \cdot 1 \cdot -2 = -8.$$

The factorial function of order n has n factors. Nevertheless, it is very easy to find its difference. For

$$\Delta x^{(n)} = (x + h)^{(n)} - x^{(n)}$$
$$= (x + h)(x)(x - h) \cdots [x - (n - 2)h]$$
$$- x(x - h)(x - 2h) \cdots [x - (n - 1)h].$$

Both terms on the right have the common factor $x^{(n-1)}$. Therefore

$$\Delta x^{(n)} = x^{(n-1)}\{(x + h) - [x - (n - 1)h]\}$$
$$= nhx^{(n-1)}.$$

This derivation, since it involves $x^{(n-1)}$, is valid for any $n > 1$, but not for $n = 1$, since $x^{(0)}$ is undefined. But $x^{(1)} = x$ and $\Delta x = h$ so if we extend the definition of the factorial function by the requirement

(1.24) $$x^{(0)} = 1,$$

we can summarize our result by writing the following *formula for the difference of factorial functions:*

(1.25) $$\Delta x^{(n)} = nhx^{(n-1)} \qquad n = 1, 2, 3, \cdots.$$

Note that except for a constant factor, the difference of a factorial function is another factorial function of order 1 lower than the original. This makes repeated differencing of $x^{(n)}$ easy to perform. For example,

$$\Delta^2 x^{(n)} = \Delta(\Delta x^{(n)}) = \Delta(nhx^{(n-1)})$$
$$= nh\Delta x^{(n-1)}$$
$$= nh \cdot (n - 1)hx^{(n-2)}$$
$$= n(n - 1)h^2 x^{(n-2)}.$$

We may similarly show that

$$\Delta^3 x^{(n)} = n(n - 1)(n - 2)h^3 x^{(n-3)}$$

and repeating the Δ operation n times,

$$\Delta^n x^{(n)} = n(n - 1)(n - 2) \cdots 3 \cdot 2 \cdot 1 h^n.$$

The number $1 \cdot 2 \cdot 3 \cdots (n - 2)(n - 1)n$ in this result is the product of the first n positive integers and arises so often that a special symbol is introduced for it.

DEFINITION 1.7. *The product of the first n positive integers is denoted by $n!$ (read "n factorial"), i.e.,*

(1.26) $$n! = 1 \cdot 2 \cdot 3 \cdots (n - 1)n.$$

In addition, we agree to define zero factorial by

(1.27) $$0! = 1.$$

With this notation we can write

(1.28) $$\Delta^n x^{(n)} = n! h^n \qquad n = 0, 1, 2, \cdots .$$

Examples

(a) $3! = 1 \cdot 2 \cdot 3 = 6, \qquad 4! = 1 \cdot 2 \cdot 3 \cdot 4 = 24,$

$$\frac{5!}{3!} = \frac{1 \cdot 2 \cdot 3 \cdot 4 \cdot 5}{1 \cdot 2 \cdot 3} = 20, \qquad \frac{6!}{2!4!} = \frac{1 \cdot 2 \cdot 3 \cdot 4 \cdot 5 \cdot 6}{(1 \cdot 2)(1 \cdot 2 \cdot 3 \cdot 4)} = 15,$$

(b) $\Delta x^{(3)} = 3hx^{(2)} = 3hx(x - h)$

$\Delta^2 x^{(3)} = 3 \cdot 2h^2 x^{(1)} = 6h^2 x$

$\Delta^3 x^{(3)} = 3! h^3 = 6h^3$

$\Delta^n x^{(3)} = 0 \quad$ if $n = 4, 5, 6, \cdots .$

(c) Using (1.20) and (1.25) we have

$$\Delta(3 + 5x^{(1)} - 2x^{(2)} + 4x^{(3)}) = 3\Delta 1 + 5\Delta x^{(1)} - 2\Delta x^{(2)} + 4\Delta x^{(3)}$$
$$= 0 + 5h - 2 \cdot 2hx^{(1)} + 4 \cdot 3hx^{(2)}$$
$$= 5h - 4hx^{(1)} + 12hx^{(2)}.$$

One of the factors contributing to the importance of the factorial functions is the fact that any positive integral power of x can be written as a sum of factorial functions. We do not prove this here (cf. 6 of Problems 1.4) but instead give some illustrative examples. It is easy to verify that $x = x^{(1)}$ and $x^2 = x^{(2)} + hx^{(1)}$. Now let us try to write x^3 in terms of $x^{(3)}$, $x^{(2)}$, and $x^{(1)}$. The factorial $x^{(3)}$, when multiplied out, involves terms in x^3, x^2, and x, so we try to add to $x^{(3)}$ just the proper proportions of $x^{(2)}$ and $x^{(1)}$ in order to cancel the x and x^2 terms and leave just x^3. We therefore begin by assuming constants A_1 and A_2 can be found for which, identically in x,

$$x^3 = x^{(3)} + A_2 x^{(2)} + A_1 x^{(1)}.$$

To find A_1 and A_2 we put $x = h$ and $x = 2h$ in this equation and obtain (noting that $x^{(3)} = x^{(2)} = 0$ when $x = h$, and $x^{(3)} = 0$ when $x = 2h$)

$$h^3 = A_1 h$$
$$(2h)^3 = A_2 2h \cdot h + A_1 2h.$$

From the first of these we have $A_1 = h^2$ and from the second, $A_2 = 3h$. Thus

(1.29) $$x^3 = x^{(3)} + 3hx^{(2)} + h^2 x^{(1)}$$

and

$$\Delta x^3 = 3hx^{(2)} + 6h^2x^{(1)} + h^3$$
$$= 3hx(x - h) + 6h^2x + h^3$$
$$= 3hx^2 + 3h^2x + h^3,$$

the same result, of course, as in Table 1.1, p. 21.

We conclude this section by pointing out that the difference of a product of two functions can be expressed in terms of the operations E and Δ on the functions themselves. The proof of the following result is left as an exercise. Additional properties of the operators Δ and E will be considered in the next section.

THEOREM 1.4. *Let u and v be two functions. Then*

(1.30) $$\Delta[u(x)v(x)] = Eu(x) \cdot \Delta v(x) + v(x) \cdot \Delta u(x).$$

Example

$$\Delta x^4 = \Delta(x \cdot x^3), \text{ so [with } u(x) = x \text{ and } v(x) = x^3]$$
$$= Ex \, \Delta x^3 + x^3 \, \Delta x$$
$$= (x + h)(3hx^2 + 3h^2x + h^3) + x^3h$$

(1.31) $$\Delta x^4 = 4hx^3 + 6h^2x^2 + 4h^3x + h^4.$$

PROBLEMS 1.4

1. Find $\Delta y(x)$ if y is given by

(a) $y(x) = 3 + 2x + x^2$
(b) $y(x) = 6 - x^2$
(c) $y(x) = 5x(1 - x)$

2. Use (1.25) to find $\Delta y(x)$, $\Delta^2 y(x)$, $\Delta^3 y(x)$, $\cdots$, if

(a) $y(x) = 5x^{(3)}$
(b) $y(x) = 2 + 3x^{(1)} - 2x^{(2)} + 3x^{(3)} - x^{(4)}$
(c) $y(x) = 2.5x^{(2)} + 3.1x^{(3)}$

3. Prove the following properties of the operators Δ and E:

(a) $E[cy(x)] = cEy(x)$
(b) $E[c_1y_1(x) + c_2y_2(x)] = c_1Ey_1(x) + c_2Ey_2(x)$
(c) $E\left[\sum_{i=1}^{n} c_iy_i(x)\right] = \sum_{i=1}^{n} c_iEy_i(x)$
(d) $E^n[c_1y_1(x) + c_2y_2(x)] = c_1E^ny_1(x) + c_2E^ny_2(x)$ $n = 0, 1, 2, \cdots$
(e) $\Delta^n[c_1y_1(x) + c_2y_2(x)] = c_1\Delta^ny_1(x) + c_2\Delta^ny_2(x)$ $n = 0, 1, 2, \cdots$

4. Prove (1.28) by mathematical induction.

5. (a) Write x^4 as a sum of factorial functions. (*Hint:* Put $x^4 = x^{(4)} + A_3x^{(3)} + A_2x^{(2)} + A_1x^{(1)}$. Then put $x = h$, $2h$, $3h$ in turn to find $A_1 = h^3$, $A_2 = 7h^2$, $A_3 = 6h$.)   (b) Find Δx^4 using the result of (a) and check with (1.31).

6. Show that x^n can be written as a sum of factorial functions for any positive integer n. Let

$$x^n = x^{(n)} + A_{n-1}x^{(n-1)} + A_{n-2}x^{(n-2)} + \cdots + A_2 x^{(2)} + A_1 x^{(1)}.$$

Then show, without doing the calculations, that A_1 is determined by putting $x = h$, A_2 by putting $x = 2h$, etc.

7. Use the result of Problem 6 together with rules for the difference of factorial functions to prove that $\Delta^n x^n = n! h^n$.

8. Prove that if y is the polynomial of degree n given by (1.21) then $\Delta^n y(x) = n! h^n a_n$. [*Hint:* Use Theorem 1.3 to show that $\Delta^n y(x) = a_n \Delta^n x^n$ and then use the result of Problem 7.]

9. Let $y(x) = x^5$. Find $\Delta y(x)$ in the following three different ways and check that each gives the same result: (*a*) directly from Definition 1.2; (*b*) by writing x^5 as the sum of factorial functions and using (1.25); (*c*) by writing $x^5 = x^2 \cdot x^3$ and using (1.30) and Table 1.1.

10. Prove Theorem 1.4.

11. Derive the following formula for the difference of the quotient of two functions u and v:

$$\Delta \frac{u(x)}{v(x)} = \frac{v(x)\Delta u(x) - u(x)\Delta v(x)}{v(x)Ev(x)} \qquad \text{if } v(x)Ev(x) \neq 0.$$

12. Let Y (income), I (investment expenditure), S (savings), and C (consumer expenditure) be four functions whose values in period t are denoted by Y_t, I_t, S_t, and C_t respectively. Assume[12] the two relations

$$Y_t = I_t + C_t \qquad \text{and} \qquad S_t = I_t$$

and prove if

$$k = \frac{\Delta Y_t}{\Delta I_t}, \qquad s = \frac{\Delta S_t}{\Delta Y_t}, \qquad \text{and} \qquad c = \frac{\Delta C_t}{\Delta Y_t},$$

that

$$k = \frac{1}{s} \qquad \text{and} \qquad k = \frac{1}{1-c}.$$

13. Let G (government expenditure), D (government deficit), and Y (national income) be functions of time t related in such a way that

$$\Delta Y_t = \Delta G_t + 4\Delta D_t.$$

Make the following inference[13] from this relation: "Then, an extra dollar of deficit would allow a four dollar reduction in tax-financed government spending without affecting the magnitude of national income $\cdots$"

---

[12] A. H. Hansen, *A Guide to Keynes*, McGraw-Hill, New York, 1953, p. 96.

[13] R. L. Bishop, "Alternative Expansionist Fiscal Policies: A Diagrammatic Analysis," in *Income, Employment and Public Policy*, essays in honor of Alvin H. Hansen, W. W. Norton, New York, 1948, pp. 317–340.

14. Suppose P (price of securities) is related to M (money deposits in banks) and B (stock of securities in banks) in such a way that for each time t,

$$P_t = k \frac{M_t}{B_t} \qquad k \text{ a constant} > 0.$$

Suppose in a time interval Δt, taken equal to 1, B increases by an amount ΔB_t and M by an amount ΔM_t. Use the result of Problem 11 to show[14] that the corresponding change in price of securities is given by

$$\Delta P_t = k \frac{B_t \Delta M_t - M_t \Delta B_t}{B_t(B_t + \Delta B_t)}.$$

Prove from this that the price of securities rises, falls, or stays the same (in the interval Δt) according as $(\Delta M_t)/M_t$ (the relative increase in money) is greater than, less than, or equal to $(\Delta B_t)/B_t$ (the relative increase in securities).

15. Let T_t denote the number of occurrences of a certain response R up to time period t, and x_t the number of stimulus elements (of a certain class) which are conditioned to R in period t. Estes[15] shows that

$$\frac{\Delta x_t}{\Delta T_t} = s \frac{S - x_t}{S} \qquad \text{and} \qquad \frac{\Delta T_t}{\Delta t} = \frac{x_t}{Sh}$$

where h, s, and S are constants. Show that

$$\frac{\Delta x_t}{\Delta t} = \frac{s(S - x_t)x_t}{hS^2}.$$

1.5 EQUIVALENCE OF OPERATORS

Let us look back for a moment to (1.11) which we rewrite here:

(1.11′) $$\Delta y(x) = Ey(x) - Iy(x).$$

One might translate this equation by saying that the same result is obtained by applying the operator Δ to any function as by applying the operator $(E - I)$. (We make the natural definition that $E - I$ is the operator which, when applied to the function y, produces the function $Ey - Iy$.) Such pairs of operators arise fairly often and a special terminology is introduced for describing them.

DEFINITION 1.8. *Two operators, O_1 and O_2, are said to be equivalent (denoted by $O_1 \equiv O_2$) if for any function y to which O_1 and O_2 are each applicable, the functions $O_1 y$ and $O_2 y$ are equal.*

[14] K. E. Boulding, *Economic Analysis*, 3rd ed., Harper, New York, 1955, p. 380.
[15] W. K. Estes, "Toward a Statistical Theory of Learning," *Psychological Review*, 57 (1950), 94–107.

Thus (1.11′) can be written as the equivalence

(1.32) $$\Delta \equiv E - I.$$

Since (1.11′) can also be written in the form

$$Ey(x) = \Delta y(x) + Iy(x),$$

we are able to write

(1.33) $$E \equiv \Delta + I.$$

Note that (1.33) follows from (1.32) *if* we assume that we can manipulate operators in equivalence equations as we do algebraic quantities in ordinary equations. We shall not prove this here, but such manipulations are indeed possible and this fact accounts for the great usefulness of operator methods in the calculus of finite differences.

Other relations, already proved, can be rewritten in our new notation. For example, (1.9), (1.14), and (1.15) are translated by (1.34), (1.35), and (1.36) respectively.

(1.34) $$\Delta^0 \equiv I$$

(1.35) $$E^0 \equiv I$$

(1.36) $$\Delta^2 \equiv E^2 - 2E + I \qquad \Delta^3 \equiv E^3 - 3E^2 + 3E - I.$$

We shall show further on that (1.36) *could* have been obtained from (1.32) had we regarded (1.32) as an algebraic equation and first squared and then cubed both Δ and $E - I$. The reader will do well to note that this analogy between equivalence relations among operators and algebraic relations among real numbers is illustrated in all of the results which follow.

For any function y we may first apply Δ and then apply E to the resulting function Δy, i.e., we may form $E(\Delta y)$. We denote this two-step operation by $E\Delta$ so that $(E\Delta)y = E(\Delta y)$. Note that the operator written symbolically as $E\Delta$ means that we perform Δ first and E second. *The order of operations is indicated by reading from right to left.* Similarly, we may first apply E to the function y and then apply Δ to Ey, i.e., we may form $\Delta(Ey)$. This two-step operation is denoted by ΔE.

As an example, consider the factorial function y given by

$$y(x) = x^{(2)}.$$

Applying Δ first and then E:

$$\Delta y(x) = 2hx^{(1)} \qquad E\Delta y(x) = 2h(x + h).$$

In the opposite order we obtain the same final result:

$$Ey(x) = (x + h)^{(2)} \qquad \Delta Ey(x) = 2h(x + h).$$

There is no a priori reason for assuming that the order of operations is always immaterial; nevertheless, this is the case for the operators Δ and E.

THEOREM 1.5.   *The operators Δ and E are commutative, i.e.,*

$$(1.37) \qquad\qquad \Delta E \equiv E\Delta.$$

PROOF.   We determine $(\Delta E)y(x)$ and $(E\Delta)y(x)$ and observe that they are identical.   For

$$\begin{aligned}(\Delta E)y(x) &= \Delta[Ey(x)] = \Delta y(x+h)\\ &= y(x+2h) - y(x+h),\end{aligned}$$

and

$$\begin{aligned}(E\Delta)y(x) &= E[\Delta y(x)] = E[y(x+h) - y(x)]\\ &= y(x+2h) - y(x+h).\end{aligned}$$

If x is a number, the index law for exponents tells us that $x^3 \cdot x^4 = x^4 \cdot x^3 = x^7$.   We leave to the reader the proof of the following result which shows that powers of the operators Δ and E obey the same rule of combination.

THEOREM 1.6.   *The operators Δ and E satisfy the index law, i.e., for m and n nonnegative integers,*

$$(1.38) \qquad\qquad \Delta^m \Delta^n \equiv \Delta^n \Delta^m \equiv \Delta^{m+n},$$

$$(1.39) \qquad\qquad E^m E^n \equiv E^n E^m \equiv E^{m+n}.$$

The operators Δ and E are commutative; each obeys the index law and as we have seen (Theorems 1.1 and 1.2, and 3 of Problems 1.4) also obeys the commutative law with respect to constants and the distributive law. The familiar rules for combining numbers stem from the fact that numbers obey the above fundamental laws.   The fact that Δ and E also obey these laws accounts for the possibility of using these operators as if they were algebraic symbols representing numbers.

Formula (1.36) shows that both Δ^2 and Δ^3 can be written in equivalent form in terms of the operator E.   Moreover, as we have already remarked, if we manipulate operators as numbers in algebra, we obtain both results in (1.36).   Let us detail the computations involved in the second of these. We want an operator, involving E, equivalent to Δ^3.   So we use (1.32) and cube both sides to obtain

$$\Delta^3 \equiv (E - I)^3.$$

Now, we carry out the cubing operation by using the binomial theorem:

$$\Delta^3 \equiv E^3 - 3E^2 I + 3E I^2 - I^3.$$

But (cf. 9 of Problems 1.3) $I^3 \equiv I^2 \equiv I$. Also for any operator O, we have $OI \equiv IO \equiv O$ (cf. 1 of Problems 1.5). Therefore

$$\Delta^3 \equiv E^3 - 3E^2 + 3E - I,$$

which is precisely the result in (1.36). The above reasoning does not prove (1.36) since we have not rigorously proved that algebraic manipulations performed with operators are valid. Such calculations will be used only to suggest results that will then have to be verified by other means.

In fact, we can express *any* positive integral power of Δ in terms of E. The above calculation, carried out for $n = 3$, will be imitated to produce the general result for Δ^n. We take the nth power of both sides of (1.32) to obtain

(1.40) $$\Delta^n \equiv (E - I)^n.$$

We want to apply the binomial theorem, but first we use this opportunity to rewrite the theorem in a more concise form than that given in footnote 10, p. 27. Using the notation introduced in (1.26) and (1.27), the new symbol $\binom{n}{k}$, for k and n nonnegative integers ($k \leq n$), is defined by

(1.41) $$\binom{n}{k} = \frac{n!}{k!(n - k)!}.$$

By dividing $n!$ by $(n - k)!$ we find

$$\binom{n}{k} = \frac{n(n - 1)(n - 2) \cdots (n - k + 1)}{k!}.$$

Rewriting the formula in footnote 10 using this new notation, we have the *alternate form of the binomial formula:*

(1.42) $$(A + B)^n = \sum_{k=0}^{n} \binom{n}{k} A^k B^{n-k}.$$

It is for this reason that the numbers $\binom{n}{k}$ for $k = 0, 1, 2, \cdots n$ are known as binomial coefficients.

Applying (1.42) to (1.40),

$$\Delta^n \equiv \sum_{k=0}^{n} \binom{n}{k} (-1)^{n-k} E^k I^{n-k}.$$

But $E^k I^{n-k} \equiv E^k$, so

$$\Delta^n \equiv \sum_{k=0}^{n} \binom{n}{k} (-1)^{n-k} E^k.$$

That this heuristic derivation actually has led to a true result is the content of the following theorem, whose long, but straightforward, proof (by mathematical induction) we omit.

THEOREM 1.7. *If n is a positive integer, then*

$$(1.43) \qquad \Delta^n y(x) = \sum_{k=0}^{n} \binom{n}{k} (-1)^{n-k} E^k y(x).$$

Theorem 1.7 shows that Δ^n may be written in terms of E. The inverse problem of relating E^n to the operator Δ will now be considered. We have seen in (1.16) the solution of this problem for E^2 and E^3. In the general case, we proceed formally from the fundamental equivalence (1.33) and raise both sides to the nth power. Once again we use the binomial formula (1.42) and regard the operators as if they were algebraic quantities. Then

$$E^n \equiv (\Delta + I)^n$$

$$\equiv \sum_{k=0}^{n} \binom{n}{k} \Delta^k I^{n-k}.$$

But $\Delta^k I^{n-k} \equiv \Delta^k$ so that

$$(1.44) \qquad E^n \equiv \sum_{k=0}^{n} \binom{n}{k} \Delta^k.$$

These manipulations are again fruitful since it can be proved (by mathematical induction) that (1.44) is, in fact, a valid equivalence. We omit the proof.

THEOREM 1.8. *If n is a positive integer, then*

$$(1.45) \qquad E^n y(x) = \sum_{k=0}^{n} \binom{n}{k} \Delta^k y(x).$$

Theorem 1.8 can be used to solve simple problems of polynomial interpolation. For example, suppose we are given the following information about a function y: (1) y is a polynomial of degree 3, and (2) $y(0) = -4$, $y(1) = -6$, $y(2) = -8$, $y(3) = -4$. We require the value $y(4)$. The information in item (2) enables us to construct the *difference table*,

TABLE 1.2

x	$y(x)$	$\Delta y(x)$	$\Delta^2 y(x)$	$\Delta^3 y(x)$
0	-4	-2	0	6
1	-6	-2	6	
2	-8	4		
3	-4			
4	?			

Table 1.2. The first two columns list the x-values and the corresponding prescribed y-values; a question mark indicates the y-value to be found. We have added columns listing the values of Δy, $\Delta^2 y$, and $\Delta^3 y$ (with a difference interval equal to 1).

To calculate these entries is an easy task. For example, we compute

$$\Delta y(0) = y(1) - y(0) = -6 - (-4) = -2$$
$$\Delta y(1) = y(2) - y(1) = -8 - (-6) = -2, \text{ etc.}$$

and the column of second differences is obtained from the first differences:

$$\Delta^2 y(0) = \Delta y(1) - \Delta y(0) = -2 - (-2) = 0, \text{ etc.}$$

Finally, we have

$$\Delta^3 y(0) = \Delta^2 y(1) - \Delta^2 y(0) = 6 - 0 = 6.$$

Now we use Theorem 1.8 to write

$$y(4) = E^4 y(0) = \sum_{k=0}^{4} \binom{4}{k} \Delta^k y(0)$$

$$= y(0) + 4\Delta y(0) + 6\Delta^2 y(0) + 4\Delta^3 y(0) + \Delta^4 y(0).$$

The successive differences of y at $x = 0$ can be read from the first row of the difference table since they are the leading entries in each column. The assumption that y is a polynomial of degree 3 implies (Theorem 1.3) that $\Delta^4 y(0) = 0$. Hence

$$y(4) = -4 + 4(-2) + 6(0) + 4(6) = 12.$$

PROBLEMS 1.5

1. If O is any operator, prove that $OI \equiv IO \equiv O$.

2. If n is any positive integer, prove that $I^n \equiv I$. (Cf. 9 of Problems 1.3.)

3. If in the definition of the factorial function we put $h = 1$, then show $\binom{n}{k} = \dfrac{n^{(k)}}{k!}$, where n and k are integers and $n \geq k \geq 0$.

4. Prove the following identities for binomial coefficients (m and j integers, $m \geq j \geq 0$):

(a) $\binom{m}{0} = \binom{m}{m} = 1$

(b) $\binom{m}{j} = \binom{m}{m-j}$

(c) $\binom{m}{j-1} + \binom{m}{j} = \binom{m+1}{j}$

(d) $\dbinom{m}{0} + \dbinom{m}{1} + \dbinom{m}{2} + \cdots + \dbinom{m}{m} = 2^m$

(e) $\dbinom{m}{0} - \dbinom{m}{1} + \dbinom{m}{2} - \cdots \pm \dbinom{m}{m} = 0$

[*Hint:* To prove (d) and (e), use the binomial theorem to write $(1 + x)^m$ and then let $x = 1$ and $x = -1$ respectively.]

5. Show that from a group of N individuals, we may select $\dbinom{N}{1}$ different individuals, $\dbinom{N}{2}$ different groups of two individuals,[16] $\dbinom{N}{3}$ different groups of three, etc.   [In general, the number of *combinations* of N different objects taken k at a time is $\dbinom{N}{k}$.   This accounts for the notation C_k^N, sometimes used in place of the symbol adopted here.]

6. Compute $\Delta E y(x)$ and $E \Delta y(x)$ and thus verify Theorem 1.5 if (a) $y(x) = x$, (b) $y(x) = x^{(3)}$, (c) $y(x) = c^x$.

7. Apply (1.43) to the function y for which $y(x) = c$, a constant, and use identity (e) of Problem 4 to show that (1.43) reduces to the trivial result that if n is a positive integer, $\Delta^n c = 0$.

8. Apply (1.43) to the function y for which $y(x) = c^x$ and thus show that

$$\Delta^n c^x = c^x \sum_{k=0}^{n} \binom{n}{k} (-1)^{n-k} c^{kh}.$$

Recognize the indicated sum as the result of applying the binomial theorem to expand $(c^h - 1)^n$ so that the equation may be written

$$\Delta^n c^x = (c^h - 1)^n c^x.$$

9. Apply (1.43) to the function y for which $y(x) = x^m$, m a positive integer, and thus derive the formula

$$\Delta^n x^m = \sum_{k=0}^{n} \binom{n}{k} (-1)^{n-k} (x + kh)^m.$$

10. The numbers $\Delta^n x^m$ when x is put equal to 0 (and $h = 1$) are denoted by $\Delta^n 0^m$ and called *differences of zero*.   Using the result of Problem 9, calculate the following differences of zero:

$\Delta 0^1 = 1$	$\Delta^2 0^1 = 0$	$\Delta^3 0^1 = 0$
$\Delta 0^2 = 1$	$\Delta^2 0^2 = 2$	$\Delta^3 0^2 = 0$
$\Delta 0^3 = 1$	$\Delta^2 0^3 = 6$	$\Delta^3 0^3 = 6$

11. Show that $\Delta 0^m = 1$ for every positive integer m.   (*Hint:* Write the result of Problem 9 with $n = h = 1$ and then put $x = 0$.)

---

[16] For an application in a formula for "group cohesion" see C. H. Proctor and C. P. Loomis, "Analysis of Sociometric Data," Chap. 17 in *Research Methods in Social Relations, Part Two: Selected Techniques*, Dryden Press, New York, 1951, p. 572.

12. In the example following Theorem 1.8, show that $y(5) = 46$. Then check this result by finding the values belonging in the empty cells of the difference table, Table 1.2. (Work to the left from the last column, all of whose entries equal 6.)

1.6 INDEFINITE SUMMATION: THE OPERATOR Δ^{-1}

Up to this point we have defined Δ^n and E^n only for positive integral (and zero) values of n. Separate considerations are required for negative indices. Consider, for example, the operator Δ^{-1}. We should like to extend our definitions in such a manner that certain basic properties (such as the linearity properties of Theorems 1.1 and 1.2 and the index law of Theorem 1.6) remain valid. So we require that the operator Δ^{-1} be defined in such a way that

$$(1.46) \qquad \Delta[\Delta^{-1}y(x)] = \Delta^{1-1}y(x) = \Delta^0 y(x) = y(x).$$

Thus, *if* the index law is to be preserved, then $\Delta^{-1}y$ should be defined as a function whose first difference is the function y. For this is precisely what is stated by (1.46).

DEFINITION 1.9. *If Y is a function whose first difference is the function y, then Y is called an indefinite sum of y and denoted by $\Delta^{-1}y$, i.e.,*

$$(1.47) \qquad if\ \Delta Y(x) = y(x),\ then\ \Delta^{-1}y(x) = Y(x).$$

Finding an indefinite sum of y is the problem inverse to finding the difference of y. Instead of *starting with y* and differencing it, we rather seek another function which *results in y* when differenced. (The inverse operations of addition and subtraction, multiplication and division, squaring and taking square roots, etc., are familiar from arithmetic.) The two equations in (1.47) express the same relation between the functions y and Y; we use the first in going from Y to y via the operator Δ, the second in going from y to Y via the operator Δ^{-1}.

As an example, let us find an indefinite sum of the constant function equal to h, the difference interval. Since $\Delta x = h$, we have $\Delta^{-1}h = x$. But the function whose value at x is given by $x + 2$ also has a first difference equal to h, so that $\Delta^{-1}h = x + 2$ as well. In fact, $\Delta(x + C) = h$ for *any* constant C and therefore

$$(1.48) \qquad \Delta^{-1}h = x + C.$$

But still more is true.   For if p is any *function with period h*, i.e., a function for which[17]

(1.49) $\Delta p(x) = p(x + h) - p(x) = 0$

for all x in the domain of p, then

$$\Delta[x + p(x)] = \Delta x + \Delta p(x) = h + 0 = h,$$

so that

(1.50) $\Delta^{-1}h = x + p(x).$

Since (1.50) is true for *any* periodic function p (with period h), we see that the constant function h has infinitely many indefinite sums.   In particular, every constant function is periodic, so (1.50) includes (1.48) as a special case.   But there are also nonconstant periodic functions, such as the sine and cosine functions in trigonometry.   (Cf. 8 of Problems 1.6.)

This example illustrates the following fact: *if Y is any indefinite sum of y, then Y + p is another indefinite sum of y if p is a function with a period equal to the difference interval.*   That we have thereby included *all* indefinite sums of y follows from the following result.

THEOREM 1.9.   *If Y_1 and Y_2 are any two indefinite sums of the same function y, then their difference $(Y_1 - Y_2)$ is a function with period h, the difference interval.*

PROOF.   We must prove that the first difference of $(Y_1 - Y_2)$ is identically zero.   But

$$\Delta(Y_1 - Y_2) = \Delta Y_1 - \Delta Y_2$$
$$= y - y \text{ (by hypothesis)}$$
$$= 0.$$

Given a function y, we may choose any one of its indefinite sums, say Y, add to it an arbitrary function p of period h, and write

(1.51) $\Delta^{-1}y(x) = Y(x) + p(x).$

This equation expresses the fact, proved in Theorem 1.9, that if Z is a function for which $\Delta Z = y$, then there is a periodic function p such that $Z = Y + p$.

---

[17] If h is any number for which (1.49) is true for all x, then h is properly referred to as *a* period of the function p.   *The* period of p is then defined as the smallest such h. For our purposes, it suffices to consider h as any period of p.

In view of (1.47), a formula concerning indefinite sums corresponds to every formula for differencing. For example, from (1.25)

$$\Delta \frac{x^{(2)}}{2h} = x^{(1)},$$

so we have

$$\Delta^{-1}x^{(1)} = \frac{x^{(2)}}{2h} + p(x).$$

Similarly, since

$$\Delta \frac{x^{(3)}}{3h} = x^{(2)},$$

we have

$$\Delta^{-1}x^{(2)} = \frac{x^{(3)}}{3h} + p(x),$$

and, in general, (1.25) yields a formula for the *indefinite sum of any factorial function:*

$$(1.52) \qquad \Delta^{-1}x^{(n)} = \frac{x^{(n+1)}}{(n+1)h} + p(x) \qquad n = 0, 1, 2, \cdots.$$

Before considering formulas for the indefinite sum of other special functions, we prove some general results.

THEOREM 1.10. *If Y_1 and Y_2 are indefinite sums of the functions y_1 and y_2 respectively, and if c_1 and c_2 are arbitrary constants, then $c_1 Y_1 + c_2 Y_2$ is an indefinite sum of the function $c_1 y_1 + c_2 y_2$, i.e.,*

$$(1.53) \qquad \Delta^{-1}(c_1 y_1 + c_2 y_2) = c_1 \Delta^{-1}y_1 + c_2 \Delta^{-1}y_2.$$

PROOF. We are given $\Delta Y_1 = y_1$ and $\Delta Y_2 = y_2$ and must prove that the difference of $c_1 Y_1 + c_2 Y_2$ is equal to $c_1 y_1 + c_2 y_2$. We use (1.19) to obtain

$$\Delta(c_1 Y_1 + c_2 Y_2) = c_1 \Delta Y_1 + c_2 \Delta Y_2$$
$$= c_1 y_1 + c_2 y_2.$$

Formula (1.53) may be extended to a linear combination of n functions by using (1.20′), the corresponding result for the difference operator. We obtain

$$(1.54) \qquad \Delta^{-1} \sum_{k=1}^{n} c_k y_k = \sum_{k=1}^{n} c_k \Delta^{-1}y_k.$$

Example

To find $\Delta^{-1}x^3$ we first write x^3 as a sum of factorials. From (1.29) we have

$$x^3 = x^{(3)} + 3hx^{(2)} + h^2 x^{(1)}$$

so that, using (1.54),

$$\Delta^{-1}x^3 = \Delta^{-1}x^{(3)} + 3h\Delta^{-1}x^{(2)} + h^2\Delta^{-1}x^{(1)}.$$

But (1.52) may now be used to find an indefinite sum of each factorial function:

$$\Delta^{-1}x^3 = \frac{x^{(4)}}{4h} + p_1(x) + x^{(3)} + p_2(x) + \frac{hx^{(2)}}{2} + p_3(x)$$

$$= \frac{x^{(4)}}{4h} + x^{(3)} + \frac{hx^{(2)}}{2} + p(x),$$

where we have written the single periodic function p in place of the sum $p_1 + p_2 + p_3$ since the sum of a finite number of functions of period h is a function of period h (cf. 5 and 6 of Problems 1.6).

We may define the operator E^{-1} as well as Δ^{-1}. In fact, we define E^{-1} as the operator inverse to E.

DEFINITION 1.10. *If Y is a function for which EY is the function y, then we write $E^{-1}y = Y$, i.e.,*

$$(1.55) \qquad \text{if } EY(x) = y(x), \text{ then } E^{-1}y(x) = Y(x).$$

The operator E^{-1} is less interesting (and less useful) than Δ^{-1}, for directly from (1.55) we see that

$$(1.56) \qquad E^{-1}y(x) = y(x - h).$$

This is due to the fact that $Ey(x - h) = y(x)$. Therefore, if a function y is given and we are asked to find another function Y such that $EY = y$, this problem is completely and trivially solved by (1.56).

In order to complete the picture, we merely mention here that the operators $\Delta^{-2}, \Delta^{-3}, \cdots, E^{-2}, E^{-3}, \cdots$ may also be defined. In fact, for any positive integer n, we define $\Delta^{-n}y$ as a function whose nth difference is y. Similarly, $E^{-n}y$ is a function for which $E^n(E^{-n}y) = y$. Noting that $E^ny(x - nh) = y(x)$, we have

$$(1.57) \qquad E^{-n}y(x) = y(x - nh).$$

PROBLEMS 1.6

1. Write x^2 as a sum of factorials and thus show that

$$\Delta^{-1}x^2 = \frac{x^{(3)}}{3h} + \frac{x^{(2)}}{2} + p(x).$$

2. Show similarly (cf. 5 of Problems 1.4) that

$$\Delta^{-1}x^4 = \frac{x^{(5)}}{5h} + \frac{3x^{(4)}}{2} + \frac{7hx^{(3)}}{3} + \frac{h^2x^{(2)}}{2} + p(x).$$

3. Show that, if $c \neq 1$,

$$\Delta^{-1}c^x = \frac{c^x}{c^h - 1} + p(x).$$

4. If y is a function with period h, show that

$$\Delta^{-1}y(x) = \frac{xy(x)}{h} + p(x).$$

5. If p_1 and p_2 are each functions with period h, prove, directly from the definition of a periodic function, that $p_1 + p_2$ is also a function with period h.

6. Prove that the sum of any finite number of functions of period h is a function of period h.

7. Use Theorem 1.9 to show that if $\Delta y = 0$, then y is a function with period h.

8. (a) Let y be the sine function; i.e., the value of y for any real number x is given by $y(x) = \sin x$. Using the trigonometric identity

$$\sin (A + B) = \sin A \cos B + \cos A \sin B$$

with $A = x$ and $B = 2\pi$, show that y is a function with period 2π. (Recall that $\sin 2\pi = 0$, $\cos 2\pi = 1$.)

(b) Let y be the cosine function with $y(x) = \cos x$. Using the identity

$$\cos (A + B) = \cos A \cos B - \sin A \sin B$$

with $A = x$ and $B = 2\pi$, show that y is a function with period 2π.

9. Let y be the function given by the equation $y(x) = x$. Show that $\Delta\Delta^{-1}y(x) = x$ but $\Delta^{-1}\Delta y(x) = x + p(x)$ where $p(x)$ is any function with period h. (This means that the index law of Theorem 1.6 cannot be extended to negative exponents.)

10. Find a function for which $\Delta^{-1}Ey(x)$ and $E\Delta^{-1}y(x)$ are unequal. Compare with Theorem 1.5.

11. Use (1.30) to prove the following formula for *summation by parts*:

$$\Delta^{-1}[v(x)\Delta u(x)] = u(x)v(x) - \Delta^{-1}[Eu(x) \cdot \Delta v(x)].$$

12. Using the formula for summation by parts, show that

$$\Delta^{-1}(x2^x) = \frac{x2^x}{2^h - 1} - \frac{h2^{x+h}}{(2^h - 1)^2} + p(x).$$

[*Hint:* Let $v(x) = x$, $\Delta u(x) = 2^x$. Then use the result of Problem 3 to find $u(x)$.]

13. (Application to finding sums of finite series.) Prove the following theorem: *Let y be a given function with domain the set of positive integers and let Y be any indefinite sum of y. Then*

$$\sum_{k=1}^{n} y(k) = Y(n + 1) - Y(1).$$

[*Hint:* By hypothesis, $\Delta Y(k) = y(k)$ for any positive integer k. Write this for $k = 1, 2, 3, \cdots, n$ and add the resulting equations.]

14. By applying the theorem of Problem 13, derive each of the following sum formulas:

(a) $\displaystyle\sum_{k=1}^{n} k = \frac{n(n+1)}{2}.$

(b) $\displaystyle\sum_{k=1}^{n} ar^{k-1} = \frac{a - ar^n}{1 - r}$ $(r \neq 1)$

(c) $\displaystyle\sum_{k=1}^{n} (2k - 1) = n^2$

[*Hints:* (a) $y(k) = k$ so take $Y(k) = k^{(2)}/2$. (b) $y(k) = ar^{k-1}$; use the result of Problem 3 to find $Y(k) = (ar^{k-1})/(r - 1)$ if $r \neq 1$. (c) $y(k) = 2k - 1$ so choose $Y(k) = k^2 - 2k$.]

15. Write the sum $a + ar + ar^2 + \cdots + ar^{n-1}$ using the summation symbol and thus observe that the result of Problem 14(b) can be stated as follows: the sum of the first n terms of a geometric progression with first term a and common ratio r is equal to $(a - ar^n)/(1 - r)$ if $r \neq 1$. Show also that if $r = 1$, the sum equals na.

*1.7 ANALOGIES BETWEEN THE DIFFERENCE AND DIFFERENTIAL CALCULUS

The fact that the derivative of a function is defined as a limit of a difference quotient accounts for the many interesting analogies between the calculus of finite differences and the differential calculus. Let us review this definition as a preliminary to the exploration of some of these analogies.

A function y being given, the new function Dy, whose value at x is

(1.58) $$Dy(x) = \lim_{h \to 0} \frac{y(x + h) - y(x)}{h} = \lim_{h \to 0} \frac{\Delta y(x)}{h},$$

(whenever this limit exists) is called the *derivative* of y. D is the differentiation operator which, when applied to a function, produces the derivative of that function. The number $[\Delta y(x)]/h$ is the slope of the straight line connecting the points on the graph of y at x and $x + h$; the number $Dy(x)$ is the slope of the tangent line at x. For example, we know that

$$\Delta x^2 = 2xh + h^2.$$

---

* Starred sections treat special topics requiring some knowledge of the differential and/or integral calculus. The material they contain is neither used nor referred to in the remainder of the text.

Therefore, by (1.58)

$$Dx^2 = \lim_{h \to 0} \left(\frac{2xh + h^2}{h} \right) = \lim_{h \to 0} (2x + h) = 2x.$$

In operator notation, successive differentiations of a function are indicated by successive applications of the operator D; i.e., the second derivative of y is denoted by D^2y and defined as the function $D(Dy)$, the third derivative is denoted by D^3y, etc.

Consider now the operation inverse to differentiation. If a function Y is such that $DY = y$, then Y is said to be an *antiderivative* (or primitive) of y. To be consistent with our notation for the inverse of the difference operator (which we denoted by Δ^{-1}), we should denote the inverse of differentiation by D^{-1} and write $Y = D^{-1}y$. Although this notation is useful in many contexts, it is more usual to write $Y = \int y$ and call Y an *indefinite integral* of the function y. Recall that there are infinitely many indefinite integrals of a function y, since if $DY = y$ then $D(Y + C) = y$, for any constant C.

TABLE 1.3

Difference Calculus	*Differential Calculus*
1. $\Delta y(x) = y(x + h) - y(x)$	1'. $Dy(x) = \lim_{h \to 0} \dfrac{\Delta y(x)}{h}$
2. $\Delta^n y = \Delta(\Delta^{n-1}y)$ $n = 1, 2, \cdots$	2'. $D^n y = D(D^{n-1}y)$ $n = 1, 2, \cdots$
3. $\Delta(cy) = c\Delta y$	3'. $D(cy) = cDy$
4. $\Delta(c_1y_1 + c_2y_2) = c_1\Delta y_1 + c_2\Delta y_2$	4'. $D(c_1y_1 + c_2y_2) = c_1Dy_1 + c_2Dy_2$
5. If y is a polynomial of degree n, then $\Delta^n y$ is constant and higher order differences are zero	5'. If y is a polynomial of degree n, then $D^n y$ is constant and higher order derivatives are zero
6. $\Delta x^{(n)} = nhx^{(n-1)}$	6'. $Dx^n = nx^{n-1}$
7. $\Delta(u \cdot v) = (Eu)\Delta v + v\Delta u$	7'. $D(u \cdot v) = uDv + vDu$
8. $\Delta \left(\dfrac{u}{v} \right) = \dfrac{v\Delta u - u\Delta v}{vEv}$	8'. $D \left(\dfrac{u}{v} \right) = \dfrac{vDu - uDv}{v^2}$
9. If $\Delta Y = y$, then $\Delta^{-1}y = Y + p$, where p is a function with period h	9'. If $DY = y$, then $\int y = Y + C$, where C is a constant

Using this notation, we list in Table 1.3 some familiar results from the differential calculus together with the analogous formulas from the difference calculus. Some of these results in the differential calculus may be easily proved from their counterparts in the difference calculus with the aid of some elementary properties of limits.

For example, a proof of 3′ proceeds as follows:

$$D(cy) = \lim_{h \to 0} \frac{\Delta(cy)}{h} \qquad \text{(from 1′)}$$

$$= \lim_{h \to 0} \frac{c\Delta y}{h} \qquad \text{(from 3)}$$

$$= c \lim_{h \to 0} \frac{\Delta y}{h} \qquad \text{(limit of constant times function equals constant times limit of function)}$$

$$= cDy \qquad \text{(from 1′)}.$$

Property 4′ may be similarly proved. In addition, the formulas for differentiation of products and quotients of functions follow, under suitable restrictions, from the corresponding results 7 and 8. For example, to derive 7′ we note (using 1′ and 7)

(1.59)
$$D(u \cdot v) = \lim_{h \to 0} \frac{\Delta(u \cdot v)}{h} = \lim_{h \to 0} \frac{(Eu)\Delta v + v\Delta u}{h}$$

$$= \lim_{h \to 0} \left[(Eu) \frac{\Delta v}{h} + v \frac{\Delta u}{h} \right]$$

$$= \lim_{h \to 0} (Eu) \frac{\Delta v}{h} + \lim_{h \to 0} v \frac{\Delta u}{h}$$

$$= \lim_{h \to 0} (Eu) \lim_{h \to 0} \frac{\Delta v}{h} + \lim_{h \to 0} v \lim_{h \to 0} \frac{\Delta u}{h},$$

the final two equalities following from the fact that the limit of the sum (product) of two functions is the sum (product) of the limits of the functions.

To continue the derivation of 7′, we assume u and v are differentiable functions. It then follows that u is continuous, i.e., that $u(x + h) \to u(x)$ as $h \to 0$. This continuity condition is expressed in finite difference operator notation by requiring that $Eu \to u$ as $h \to 0$. In (1.59) we now have only to evaluate the indicated limits to obtain 7′.

If $y(x)$ denotes the distance moved (along a straight line) by a particle in time x, then the difference quotient (writing Δx in place of h)

(1.60)
$$\frac{\Delta y(x)}{\Delta x}$$

is, for the time interval from x to $x + \Delta x$, the ratio of the distance moved to the time required to move this distance. That is, (1.60) gives the *average speed* in the time interval from x to $x + \Delta x$. The limit (as

$\Delta x \to 0$) of the average speed is defined as the *instantaneous speed* at time x. From (1.60), the instantaneous speed at x is given by $Dy(x)$.

Similar distinctions between *average* rates of change (given by difference quotients) and *instantaneous* rates of change (given by derivatives) are made in other fields.   For example, if $y(x)$ is the total cost of producing x units of a commodity, then (1.60) is the average rate of change of cost with respect to the number of units produced, computed for an output change from x to $x + \Delta x$ units.   The derivative $Dy(x)$ is the corresponding instantaneous rate of change.   It is called the *marginal cost* at output x. Similarly, if $y(x)$ is the total revenue at output x, then $Dy(x)$ is the *marginal revenue* at output x.

The concept of *elasticity* in economics is defined in a similar manner. The (instantaneous) elasticity of y with respect to x is a function e whose value at x is given by

$$e(x) = \lim_{\Delta x \to 0} \frac{\Delta y(x)/y(x)}{\Delta x/x};$$

i.e., e is the limit of the ratio of the *relative* increment in y to the *relative* increment in x.   Since this ratio is equal to

$$\frac{x}{y(x)} \cdot \frac{\Delta y(x)}{\Delta x},$$

we see, from (1.58), that

$$e(x) = \frac{x}{y(x)} \, Dy(x).$$

[The reader may want to show that if y is proportional to x, i.e., $y(x) = kx$, then $e(x) = +1$; if y is inversely proportional to x, i.e., $y(x) = k/x$, then $e(x) = -1$; in general, e will not be a constant function, as in these extreme examples, but will vary in value as x varies.]

Two other examples of average and instantaneous rates: If $y(x)$ represents the proportion of a population favoring war at time x, Richardson (see 16 of Problems 1.1) studies the function Dy, whose value at x is the instantaneous rate of conversion to the prowar opinion at time x.   Finally, if one is studying responses to some stimulus and $y(x)$ is the number of responses up to time x, then $\Delta y(x)/\Delta x$ is the average response rate in the time interval from x to $x + \Delta x$, and $Dy(x)$ is the instantaneous rate of responding at time x.

In view of the limit relation (1.58), it may be expected that, in general, if the difference interval h is small, then the derivative $Dy(x)$ and the difference quotient $\Delta y(x)/h$ will be approximately equal.   This fact is used in many applications (see Section 2.12).

2

Difference Equations

2.1 BASIC DEFINITIONS

In this chapter we begin our study of difference equations with a detailed analysis of one simple type of equation and its applications in the social sciences. This analysis presupposes a clear understanding of what is meant when we speak of a difference equation and "solving" such an equation. It is to these basic matters that we devote the first two sections of this chapter.

DEFINITION 2.1. *An equation relating the values of a function y and one or more of its differences Δy, $\Delta^2 y$, $\cdots$ for each x-value of some set of numbers S (for which each of these functions is defined) is called a difference equation over the set S.*

A difference equation is a relation involving differences. It is important to keep in mind that a particular function y being given, a difference equation over the set S may be a true relation (if, for this y, the equation is a true statement for each x-value in S) or it may be a false relation (if, for this y, the equation is false for one or more of the x-values in S). We are familiar with a simpler, but analogous, situation in algebra. Thus, a particular x-value being given, the algebraic equation $2x - 4 = 0$, for example, may be true (if $x = 2$) or it may be false (if $x \neq 2$).

In algebra, one studies various methods for solving algebraic equations, i.e., for finding all those x-values for which a given equation is true. In the theory of difference equations, we are similarly led to the following important problem: given a difference equation over some set S, find all those functions y for which this relation is true. We begin our study of this problem in the next section. But first we give some examples of difference equations and take up certain other preliminary matters.

If we consider functions y defined for all real numbers x, the following equations are examples of difference equations over the set S of real numbers:

(1) $$\Delta y(x) + 3y(x) = 0$$

(2) $$\Delta^2 y(x) + 2\Delta y(x) + y(x) = 0$$

(3) $$\Delta^2 y(x) - xy(x) = 2x + 7$$

(4) $$y(x)\Delta^3 y(x) = \tfrac{1}{2}$$

(5) $$[\Delta y(x)]^2 + [y(x)]^2 = -1.$$

It is important to recognize that these equations may also be thought of as difference equations over some set other than that of all real numbers. Since y is defined for all real numbers x, it is a fortiori defined for all positive integral values of x. If we use $h = 1$ (or any other positive integer), then $\Delta y(x)$, $\Delta^2 y(x)$, $\cdots$ will also be meaningful, for under these circumstances they will involve values of y at the *integers* $x + h$, $x + 2h$, $\cdots$. Therefore equations (1)–(5) may be regarded as difference equations over the set of positive integers. One cannot tell by looking at equations (1)–(5) whether they are difference equations over the set of all real numbers, over the set of all positive integers, or over some other set. This information must be explicitly given. To say, as in Definition 2.1, that we have a difference equation over some set S tells us *two* things: (1) the relation among the values of y, Δy, $\Delta^2 y$, $\cdots$ given by the equation, and (2) the set of values, denoted by S, for which this relation is said to hold.

As a matter of fact, in practically all of our work we shall need to consider difference equations over only special sets of values of x characterized as follows: the set of x-values contains some number x_0 and either a finite number or all of the successive values $x_0 + h$, $x_0 + 2h$, $x_0 + 3h$, $\cdots$. The set of values is *a finite or infinite set of equally spaced x-values* (with x_0 denoting the starting value and h the number to be added to each x-value to get the next x-value). For a discussion of the extent to which such discrete sets of x-values arise in social science problems, see the beginning paragraphs of Chapter 0.

It is possible, *without loss of generality*, to go one step further in simplifying the set S of values for which the difference equations we shall study are defined. We shall assume that x_0 is a nonnegative integer and that h is equal to 1. Then the set S consists of either *a finite or infinite set of successive integers*. (It will often be convenient to have this set start with 0 although this is not necessary.)

Before proving that if S consists of some finite or infinite set of equally spaced x-values, then, *without loss of generality*, we can assume this set to be a set of consecutive integers, we make the result plausible with an example. Suppose that we are given a function defined at the end of the 10-year intervals starting with 1900, i.e., the function, which may denote total population in some country, is known for 1900, 1910, 1920, $\cdot$ $\cdot$ $\cdot$. Suppose we want to reduce this set of x-values to the set of consecutive integers 1, 2, 3, $\cdot$ $\cdot$ $\cdot$. We first subtract 1890 from each date to obtain the sequence 10, 20, 30, $\cdot$ $\cdot$ $\cdot$, and then we have only to divide each of these values by 10 to obtain the desired set. This is just the mathematical way of saying that we may think of 1900 as the first period, 1910 as the second period, etc. If we know the year in the original sequence, we obtain the period by subtracting by 1890 and dividing by 10. If we know the period in the new sequence, we obtain the original year by multiplying by 10 and adding 1890.

To prove the general result, suppose we are given some set of equally spaced x-values beginning with any number (not necessarily an integer) and with spacing h (not necessarily an integer), i.e., we are given the set of x-values

$$(2.1) \qquad x: x_0, x_0 + h, x_0 + 2h, \cdot \cdot \cdot.$$

Now, if we let the new variable k be related to x by the equation

$$(2.2) \qquad k = \frac{x - (x_0 - ah)}{h} \qquad a \text{ any nonnegative integer,}$$

then $k = a$ if $x = x_0$, $k = a + 1$ if $x = x_0 + h$, and in general the set of x-values (2.1) is transformed into the set of consecutive integers

$$k: a, a + 1, a + 2, \cdot \cdot \cdot.$$

We lose no generality in thinking of our functions as defined for this set of k-values, for each k-value is just a code number for one and only one of the original x-values. Given any k-value, we find the corresponding x by solving (2.2) for x:

$$(2.3) \qquad x = (x_0 - ah) + kh.$$

To emphasize the restriction of the set S to consecutive integers, we shall use the symbol k rather than x to denote a number in the domain of the functions related by the difference equation, and write y_k rather than $y(k)$ for the value of y at k. *All difference operators are taken with a difference interval equal to* 1 since we have consecutive values of k. Now equations (1)–(5) are written as follows:

$$(1') \qquad \Delta y_k + 3y_k = 0$$

$$(2') \qquad \Delta^2 y_k + 2\Delta y_k + y_k = 0$$

$$(3') \qquad \Delta^2 y_k - ky_k = 2k + 7$$

$$(4') \qquad y_k \Delta^3 y_k = \tfrac{1}{2}$$

$$(5') \qquad (\Delta y_k)^2 + (y_k)^2 = -1.$$

A difference equation written in this form will be understood as being defined over some set of consecutive nonnegative integers. This set may start with any integer and may be finite or infinite. *In each case we must specify the range of values of the integer k.*

Because of Theorem 1.7 we can write any difference equation as a relation among the values of y and one or more of the functions Ey, E^2y, $\cdots$ as well as among y and Δy, Δ^2y, $\cdots$. Thus, since

$$\Delta y_k = Ey_k - y_k,$$

equation (1') may be rewritten in the form

(2.4) $$Ey_k + 2y_k = 0.$$

But we know from (1.13) that

(2.5) $$E^n y_k = y_{k+n}.$$

Therefore, we can write any difference equation as a relation among the values of y at various points of the set S for which the difference equation is defined. For example, from (2.4) and (2.5), we obtain equation (1') in the form

(1'') $$y_{k+1} + 2y_k = 0.$$

In the same way, since

$$\Delta^2 y_k = E^2 y_k - 2Ey_k + y_k,$$

we may write equation (2') in the form

(2'') $$y_{k+2} = 0.$$

Similarly, we may rewrite equations (3'), (4'), and (5'):

(3'') $$y_{k+2} - 2y_{k+1} + (1 - k)y_k = 2k + 7$$

(4'') $$y_k y_{k+3} - 3y_k y_{k+2} + 3y_k y_{k+1} - y_k^2 = \tfrac{1}{2}$$

(5'') $$(y_{k+1} - y_k)^2 + y_k^2 = -1.$$

When written in this form, as we shall do from now on, the range of values of the integer k must be specified for each equation.

Equations (1''), (2''), and (3'') are said to be linear difference equations. In general, we make the following definition:

DEFINITION 2.2. *A difference equation over a set S is linear over S if it can be written in the form*

(2.6) $$f_0(k)y_{k+n} + f_1(k)y_{k+n-1} + \cdots + f_{n-1}(k)y_{k+1} + f_n(k)y_k = g(k),$$

where $f_0, f_1, \cdots, f_{n-1}, f_n$, and g are each functions of k (but not of y_k) defined for all values of k in the set S.

Equation (1″) is of this form with $n = 1, f_0(k) = 1, f_1(k) = 2, g(k) = 0$. Equation (2″) may be written in the form (2.6) by putting $n = 2, f_0(k) = 1$, $f_1(k) = 0$, $f_2(k) = 0$, $g(k) = 0$. In these equations, the coefficients f_0, $f_1, \cdots, f_n$ are constant functions, i.e., they do not vary with k. Such equations, called *linear equations with constant coefficients*, are of fundamental importance and will form the subject matter of the next chapter. Note that equation (3″) is not of this form; although linear, it does not have constant coefficients because of $f_2(k)$ being equal to $(1 - k)$. Finally, equations (4″) and (5″) are examples of nonlinear difference equations.

Equations (1″), (2″), and (3″), although all linear equations, are different in the following sense: equation (1″) relates the value y_k with the *one* neighboring value y_{k+1}; equation (2″) relates the value y_{k+2} with *no* other values; equation (3″) relates the value of y_k with the *two* neighboring values y_{k+1} and y_{k+2}. In general, we make the following definition to distinguish these types of equations.

DEFINITION 2.3. *A linear equation over a set S is of order n over S if, when written in the form* (2.6), *both f_0 and f_n are different from zero at each point of S.*

This is merely a formal way of saying that a linear equation is of order n if it actually relates values of y for k-values (in the set S) differing by n but no more than n. The following difference equations are all linear with the order indicated:

$$y_{k+2} + 5y_{k+1} - 7y_k = 2k \qquad \text{(order 2)}$$
$$y_{k+2} + 5y_{k+1} = 2k \qquad \text{(order 1)}$$
$$y_{k+2} - 7y_k = 2k \qquad \text{(order 2)}$$
$$ky_{k+2} + 2y_{k+1} - 6y_k = 0 \qquad \text{(order 2 if } S \text{ does not contain 0)}$$
$$4^k y_{k+3} - 3^k y_{k+2} + 2^k y_{k+1} - y_k = 1 \qquad \text{(order 3).}$$

Only the first three of these equations are linear difference equations with constant coefficients.

If when written in the form (2.6), either f_0 or f_n is 0 at some point of S, then Definition 2.3 does not apply. Thus, the order of the fourth difference equation in our list is undefined if S does contain 0. Although it is possible to extend Definition 2.3 to cover these cases, it will suffice for our purposes to leave the order of such difference equations undefined.

1. Write each of the following difference equations in the same form as that of equations $(1'')-(5'')$:

 (a) $\Delta y_k = 0$
 (b) $\Delta y_k - 2y_k = 5$
 (c) $\Delta^2 y_k = 1$
 (d) $\Delta^2 y_k + 3\Delta y_k - 3y_k = k$
 (e) $\Delta^3 y_k + \Delta^2 y_k + \Delta y_k + y_k = 0$

2. The difference equations in Problem 1 are all linear with constant coefficients. Find the order of each equation. (Note that your answers do not depend on the set of k-values over which the difference equations are defined. This is not generally the case.)

2.2 SOLUTIONS OF A DIFFERENCE EQUATION

In elementary algebra, *solving* the algebraic equation

$$(2.7) \qquad\qquad 2x - 1 = 3$$

was defined to mean finding all those values of x which satisfy this equation, i.e., all values of x which, when substituted in (2.7), make the equation a true statement. Such a value of x was said to be a solution of the equation. We know that (2.7) has exactly one solution, $x = 2$. The more complicated quadratic equation

$$x^2 - 4x + 3 = 0$$

has exactly two solutions, $x = 1$ and $x = 3$. The reader should recall the fact that an equation may not have any solutions; it is possible that none of the real numbers at our disposal are able to meet the condition imposed by the equation. Such is the case with the equation

$$(2.8) \qquad\qquad x^2 = -1$$

since the square of any real number must be nonnegative. (Of course, it is possible to extend the number system to include not only the real numbers but also the so-called complex numbers. Then (2.8) has two solutions: the complex numbers $x = \sqrt{-1}$ and $x = -\sqrt{-1}$. This emphasizes the need for stating what numbers are available to us in solving an algebraic equation.)

Now suppose we are given a difference equation, say

$$(2.9) \qquad\qquad y_{k+1} - 2y_k = 0 \qquad k = 0, 1, 2, \cdots$$

(defined over the indicated set of k-values) and asked to solve this equation. We must first ask, "What is meant by a solution of a difference equation?" Let us discuss this question in the context of (2.9).

Equation (2.9) is a relation between values of a function y at the points k and $k + 1$. Is there a function, defined with domain the set of non-negative integers, which makes this equation a true statement for every one of the k-values over which the equation is defined? The answer is yes, as is shown by the function

$$(2.10) \qquad y_k = 2^k \qquad k = 0, 1, 2, \cdots .$$

For this function y we have

$$y_{k+1} - 2y_k = 2^{k+1} - 2 \cdot 2^k = 0 \qquad k = 0, 1, 2, \cdots$$

so that y satisfies the difference equation (2.9). The function y defined in (2.10) is said to be a solution of the difference equation (2.9). It is one of many solutions. Each of the functions whose value at the nonnegative integer k is given by

$$(2.11) \qquad y_k = 3 \cdot 2^k, \qquad y_k = 4 \cdot 2^k, \qquad y_k = \tfrac{1}{5} \cdot 2^k,$$

can similarly be shown to satisfy the difference equation. As a matter of fact, if C is any constant and y is given by

$$(2.12) \qquad y_k = C \cdot 2^k \qquad k = 0, 1, 2, \cdots ,$$

then y is a solution of (2.9). This follows by noting that

$$y_{k+1} - 2y_k = C \cdot 2^{k+1} - 2 \cdot C \cdot 2^k = C(2^{k+1} - 2^{k+1}) = 0,$$

so that the function y of (2.12) satisfies the difference equation (2.9). By suitably choosing the value of the arbitrary constant in the solution (2.12), we can obtain any of the other solutions in (2.11). In this sense, solution (2.12) is more general than any of the others. Any one of the functions y given in (2.10) or (2.11) is called a *particular solution* of the difference equation (2.9). The function y in (2.12), containing the arbitrary constant C, is called the *general solution* of the difference equation.

Just as in algebraic equations, difference equations may have no solution. Equation (5″) is of this character since there is no real-valued function y for which

$$(y_{k+1} - y_k)^2 + y_k{}^2 = -1.$$

DEFINITION 2.4. *A function y is a solution of a difference equation over a set S if the values of y reduce the difference equation to an identity over S, i.e., if the values of y make the difference equation a true statement for every point of S.*

If a function is a solution of a difference equation, then it is said to *satisfy* the difference equation.

Examples

(a) The function y given by

(2.13) $$y_k = 1 - \frac{2}{k} \qquad k = 1, 2, 3, \cdots$$

is a solution of the first-order difference equation

(2.14) $$(k + 1)y_{k+1} + ky_k = 2k - 3 \qquad k = 1, 2, \cdots.$$

To prove this we have only to substitute the values of y into (2.14) and see that an identity results for all positive integral values of k. From (2.13) we know that

$$y_{k+1} = 1 - \frac{2}{k + 1}$$

so (2.14) becomes

$$(k + 1)\left(1 - \frac{2}{k + 1}\right) + k\left(1 - \frac{2}{k}\right) = 2k - 3.$$

Simplifying the left-hand side of this equation, we see it is equal to

$$(k + 1) - 2 + k - 2 = 2k - 3$$

so the function y does satisfy the difference equation (2.14).

(b) The difference equation

(2.15) $$y_{k+2} - 4y_{k+1} + 4y_k = 0 \qquad k = 0, 1, 2, \cdots$$

has the solution

(2.16) $$y_k = 2^k(C_1 + C_2 k) \qquad k = 0, 1, 2, \cdots$$

for any constants C_1 and C_2.

Proof. Substitute (2.16) into (2.15) and obtain

(2.17) $$2^{k+2}[C_1 + C_2(k + 2)] - 4 \cdot 2^{k+1}[C_1 + C_2(k + 1)] + 4 \cdot 2^k(C_1 + C_2 k) = 0.$$

We need only show that this is an identity. Simplifying the left-hand side by factoring 2^{k+2} (noting that $4 \cdot 2^{k+1} = 2 \cdot 2^{k+2}$, $4 \cdot 2^k = 2^{k+2}$), we find it equal to

$$2^{k+2}\{C_1 + C_2(k + 2) - 2[C_1 + C_2(k + 1)] + C_1 + C_2 k\}.$$

Algebraic simplification shows that the quantity in braces is 0 for every $k = 0, 1, 2, \cdots$ and for every C_1 and C_2. Thus (2.17) is an identity and (2.16) satisfies (2.15).

Note that we have found infinitely many different solutions of (2.15), one for each pair of values of the constants C_1 and C_2. Suppose we are asked to find a solution of (2.15) for which

$$y_0 = 1 \qquad \text{and} \qquad y_1 = 6.$$

These so-called *initial conditions* are requirements to be met by a function y in addition to that of satisfying the difference equation. Let us see whether there is such a particular solution among all those already found. If we put $k = 0$ and $k = 1$ in (2.16) we find

$$y_0 = C_1 \qquad \text{and} \qquad y_1 = 2(C_1 + C_2),$$

so we require the constants C_1 and C_2 to satisfy the equations

$$C_1 = 1 \quad \text{and} \quad 2(C_1 + C_2) = 6.$$

These two simultaneous algebraic equations for C_1 and C_2 are easy to solve. We find

$$C_1 = 1 \quad \text{and} \quad C_2 = 2.$$

Therefore the function given by

(2.18) $$y_k = 2^k(1 + 2k)$$

is a particular solution of the difference equation (2.15) satisfying the prescribed initial conditions.

Consider another example. We have already found that the function

(2.19) $$y_k = C \cdot 2^k \qquad k = 0, 1, 2, \cdots$$

satisfies the difference equation

(2.20) $$y_{k+1} - 2y_k = 0 \qquad k = 0, 1, 2, \cdots$$

for every value of the constant C. If we want to find a solution of (2.20) satisfying the *initial condition*

(2.21) $$y_0 = 3,$$

one way to proceed would be to inquire whether among all the solutions given by (2.19) there is a solution satisfying this additional condition. From (2.19) we find

$$y_0 = C \cdot 2^0 = C,$$

so if we choose $C = 3$, we have the solution

(2.22) $$y_k = 3 \cdot 2^k \qquad k = 0, 1, 2, \cdots$$

satisfying *both* the difference equation (2.20) *and* the initial condition (2.21).

It will follow as a consequence of the theorem proved in the next section that (2.22) is the one and only solution of the difference equation (2.20) which in addition satisfies the initial condition (2.21). But we may easily prove this result directly in this special case. For the knowledge of the value y_0 determines a unique value y_1, which in turn determines a unique value y_2, etc. The function y is thus uniquely determined. The difference equation has infinitely many solutions, but only one which satisfies the given initial condition.

As we have seen, in general a difference equation may be expected to have many solutions. Solving a difference equation means finding all solutions. However, in most of the applications of difference equations to applied problems (in both the social and natural sciences) we are often required to find the solution or solutions of a difference equation satisfying one or more subsidiary conditions. The standard method of solving such a problem is first to solve the difference equation, i.e., find *all* the solutions of the difference equation, and then to select the particular solution or solutions satisfying the subsidiary conditions. This is the procedure we shall follow when we discuss linear difference equations with constant coefficients of order 2 or more. In the next section, we prove a fundamental theorem which guarantees both the existence and uniqueness of particular solutions of linear difference equations (of any order) satisfying certain subsidiary conditions.

PROBLEMS 2.2

1. A difference equation over the set $0, 1, 2, \cdots$ and a function are given. In each case, show that the function is a solution of the difference equation. (C, C_1, and C_2 denote arbitrary constants.)

(a) $y_{k+1} - y_k = 0$ $\qquad\qquad$ $y_k = 5$

(b) $y_{k+1} - y_k = 0$ $\qquad\qquad$ $y_k = C$

(c) $y_{k+1} - y_k = 1$ $\qquad\qquad$ $y_k = k$

(d) $y_{k+1} - y_k = 1$ $\qquad\qquad$ $y_k = k + C$

(e) $y_{k+1} - y_k = k$ $\qquad\qquad$ $y_k = \dfrac{k(k-1)}{2}$

(f) $y_{k+1} - y_k = k$ $\qquad\qquad$ $y_k = \dfrac{k(k-1)}{2} + C$

(g) $y_{k+2} - 3y_{k+1} + 2y_k = 0$ $\qquad$ $y_k = C_1 + C_2 \cdot 2^k$

(h) $y_{k+2} - 3y_{k+1} + 2y_k = 1$ $\qquad$ $y_k = C_1 + C_2 \cdot 2^k - k$

(i) $y_{k+2} - y_k = 0$ $\qquad\qquad$ $y_k = C_1 + C_2(-1)^k$

(j) $y_{k+1} = \dfrac{y_k}{1 + y_k}$ $\qquad\qquad$ $y_k = \dfrac{C}{1 + Ck}$

2. For the difference equations in Problem 1(d), (f), and (j), find the particular solutions satisfying the initial condition $y_0 = 1$. (*Hint:* Use the solution and the initial condition to find the value of C for the required particular solution.)

3. For the difference equations in Problem 1(g), (h), and (i), find the particular solutions satisfying the two initial conditions $y_0 = 1$ and $y_1 = 2$. (*Hint:* Use the given solution and the initial conditions to obtain two simultaneous equations for the constants C_1 and C_2.)

4. Consider the difference equation

$$D_{k+1} - 2D_k + D_{k-1} = 2 \qquad k = 1, 2, \cdots, (n-1)$$

defined over the indicated finite set of k-values where n is a positive integer greater than 1. When $k = 1$, the equation involves the value D_0 and when $k = (n-1)$, it involves D_n. Therefore, if D is a solution of this difference equation, D must be defined for $k = 0, 1, 2, \cdots, n$. Show that

$$D_k = k(n - k) \qquad k = 0, 1, 2, \cdots, n$$

is a solution of both this difference equation and the *boundary conditions*

$$D_0 = D_n = 0.$$

5. Starting from the solution of Problem 1(c), we may write

$$y_k = k + C \qquad \text{and} \qquad y_{k+1} = k + 1 + C.$$

Now eliminate C from these two equations to obtain

$$y_{k+1} = y_k + 1,$$

the difference equation satisfied by the solution with which we started. In this way, starting with the solutions of Problem 1(f) and (j), recover the corresponding difference equations.

6. Use the method of Problem 5 to obtain the difference equations of Problem 1(g), (h), and (i). (*Hint:* Use the solutions to write y_k, y_{k+1}, and y_{k+2} and then use these three equations to eliminate the two constants C_1 and C_2.)

2.3 AN EXISTENCE AND UNIQUENESS THEOREM

Since some difference equations have infinitely many solutions whereas others have no solutions at all, it is important (and comforting) to know that for the class of *linear* difference equations we can always find at least one solution and, under certain conditions, only one solution. Theorems establishing such results are referred to as *existence* and *uniqueness* theorems. They are of fundamental importance in many branches of mathematics.

Before stating and proving the theorem for the linear difference equation of order n, let us first look at the special case of order 2. The second-order linear difference equation has the form

$$(2.23) \quad f_0(k)y_{k+2} + f_1(k)y_{k+1} + f_2(k)y_k = g(k) \qquad k = 0, 1, 2, \cdots,$$

and we know the coefficients of y_k and y_{k+2} are never zero, i.e.,

$$(2.24) \qquad f_0(k) \neq 0, \qquad f_2(k) \neq 0 \qquad k = 0, 1, 2, \cdots.$$

With $k = 0$, (2.23) reads

$$f_0(0)y_2 + f_1(0)y_1 + f_2(0)y_0 = g(0),$$

so knowing any one of the values y_0, y_1, or y_2 does not enable us to find the other two. But if we know two consecutive values, y_0 and y_1, we can find y_2. For

$$f_0(0)y_2 = g(0) - f_1(0)y_1 - f_2(0)y_0$$

and in view of (2.24) we may divide[1] by $f_0(0)$ to obtain y_2. Now we use

---

[1] Division by 0 is prohibited. If $c \neq 0$, then $cx = cy$ may be divided by c to obtain $x = y$. But one cannot divide both sides of the true equation $0 \cdot 2 = 0 \cdot 3$ by 0 to obtain $2 = 3$.

the *pair* of known values y_1 and y_2 to find y_3. We have only to write the difference equation with $k = 1$ and transpose terms to get

$$f_0(1)y_3 = g(1) - f_1(1)y_2 - f_2(1)y_1.$$

Again division by $f_0(1)$ is allowed so y_3 is uniquely determined. And we can continue in this way, generating the unique solution of the second-order equation (2.23) initiated by the two values y_0 and y_1.

Any other pair of consecutive values of y would similarly suffice for the determination of a single solution. So, if y_5 and y_6 were specified, for example, we could use the difference equation to successively obtain y_4, y_3, y_2, y_1, and y_0 as well as $y_7, y_8, \cdots$.

The fact that our difference equations were assumed to be over the set of k-values $0, 1, 2, \cdots$ certainly is not crucial here. If instead we start with some integer a and either continue indefinitely or end with another integer b, i.e., if the difference equations are defined over one of the sets of integral k-values defined by

(2.25) $a \leq k$ or $a \leq k \leq b$ a, b nonnegative integers,

then again a unique solution of the second-order equation is obtained when two consecutive y-values are given. For purposes of generality, we shall formulate the following theorem, assuming the difference equation defined over either one of the sets given by (2.25). Our result will then apply to all linear difference equations over sets S of the type previously delimited for our consideration, i.e., sets of finitely or infinitely many consecutive nonnegative integers.

THEOREM 2.1.   *The linear difference equation of order n*

(2.26) $f_0(k)y_{k+n} + f_1(k)y_{k+n-1} + \cdots + f_{n-1}(k)y_{k+1} + f_n(k)y_k = g(k)$

over a set S of consecutive integer values of k has one, and only one, solution y for which values at n consecutive k-values are arbitrarily prescribed.

PROOF.   By hypothesis, S is a set of one of the types in (2.25). Suppose first that the n prescribed values of y are $y_a, y_{a+1}, \cdots, y_{a+n-1}$. We prove, using mathematical induction, that the value of y at each point of S is uniquely determined, thus proving that there is a unique solution.

The prescribed values determine the value y_{a+n} uniquely. For with $k = a$ in (2.26) we have

$$f_0(a)y_{a+n} = g(a) - f_1(a)y_{a+n-1} - \cdots - f_{n-1}(a)y_{a+1} - f_n(a)y_a$$

from which y_{a+n} is determined by division by the (nonzero) number $f_0(a)$.

Now let us make the induction hypothesis that y is known for all k-values in S up to and including y_{a+j}, where $j \geq n$. We complete the proof by showing that the following value, y_{a+j+1}, is uniquely determined.

Write the difference equation (2.26) with $k = k_1 = a + j + 1 - n$ to obtain

$$(2.27) \quad f_0(k_1)y_{a+j+1}$$
$$= g(k_1) - f_1(k_1)y_{a+j} - f_2(k_1)y_{a+j-1} - \cdots - f_n(k_1)y_{a+j-n+1}.$$

By our induction assumption, the values of y appearing on the right-hand side of this equation are known. Also, $f_0(k_1) \neq 0$, since f_0 is never zero in S. Thus division by $f_0(k_1)$ is permitted and we can find y_{a+j+1} from (2.27). We have now proved that y is uniquely determined for *all* k in S provided that the values $y_a, y_{a+1}, \cdots, y_{a+n-1}$ are prescribed.

If the first of the n consecutive prescribed values of y is y_m $(m > a)$ rather than y_a, we may first successively determine unique values for $y_{m-1}, y_{m-2}, \cdots, y_{a+1}, y_a$ and then proceed as before to show that all other values of y are uniquely determined. The proof that this may be done is also by mathematical induction. We show here that y_{m-1} is determined. The remainder of the proof may be supplied by the reader. Write the difference equation with $k = m - 1$ to obtain

$$(2.28) \quad f_n(m - 1)y_{m-1}$$
$$= g(m - 1) - f_0(m - 1)y_{m-1+n} - \cdots - f_{n-1}(m - 1)y_m.$$

Since we are supposing that $y_m, y_{m+1}, \cdots, y_{m+n-1}$ are the prescribed y values, the right-hand side of (2.28) is known. Furthermore, f_n is never zero so we may divide by $f_n(m - 1)$ to determine y_{m-1}.

In any case, whether S is the finite set $a \leq k \leq b$ or the infinite set $k \geq a$, and no matter which n consecutive values of y are prescribed, we have a unique value of y determined for every k-value in S and therefore have a unique function which satisfies the difference equation and assumes the prescribed values.

This theorem is valuable, not because it tells us how to find a solution, but because it assures us that once we have managed to find a solution of the difference equation satisfying the prescribed subsidiary conditions, we can rest assured that it is the only possible solution.

PROBLEMS 2.3

1. The difference equation

$$ky_{k+1} - y_k = 0 \qquad k = 0, 1, 2, \cdots$$

is linear but not of first order over the indicated set of k-values. (Why?) If the initial condition $y_0 = 0$ is prescribed, show that y_1 is *not* uniquely determined and there are infinitely many solutions of the difference equation with $y_0 = 0$. Show also that if the value of y at any k-value different from 0 is prescribed, there *is* a unique solution of the difference equation.

2. Consider the linear difference equation of order 2

$$y_{k+2} - ky_{k+1} - y_k = 0 \qquad k = 0, 1, 2$$

over the indicated set of three k-values. Show that there is no solution of this difference equation for which $y_0 = 0$ and $y_2 = 1$. Show also that if the prescribed values y_0 and y_2 are equal (say $y_0 = y_2 = 1$), there are infinitely many different solutions of the difference equation. Why is Theorem 2.1 not violated by these facts?

2.4 THE EQUATION $y_{k+1} = Ay_k + B$

The linear first-order difference equation has the form

$$(2.29) \qquad f_0(k)y_{k+1} + f_1(k)y_k = g(k) \qquad k = 0, 1, 2, \cdots$$

over the indicated set of k-values. The functions f_0 and f_1 are never zero so if they are constant functions, they are nonzero constants. Dividing (2.29) by $f_0(k)$ and transposing terms, we obtain

$$y_{k+1} = -\frac{f_1(k)}{f_0(k)} y_k + \frac{g(k)}{f_0(k)}.$$

If we now assume that f_0 and f_1, as well as g, are constant functions, we can write this equation in the form

$$(2.30) \qquad y_{k+1} = Ay_k + B \qquad k = 0, 1, 2, \cdots,$$

where A and B are constants and $A \neq 0$. (The constant B may be zero since g may have been identically zero in the original equation.) In this section we solve the simple difference equation (2.30).

Suppose first that y_0 is prescribed. Then (2.30), with $k = 0$, tells us that

$$(2.31) \qquad y_1 = Ay_0 + B.$$

Now, with $k = 1$, we find from the difference equation and (2.31),

$$y_2 = Ay_1 + B = A(Ay_0 + B) + B.$$

Therefore

$$(2.32) \qquad y_2 = A^2y_0 + B(1 + A).$$

We now use our knowledge of y_2 to find y_3 from the difference equation. We put $k = 2$ to obtain

$$y_3 = Ay_2 + B = A[A^2y_0 + B(1 + A)] + B.$$

Simplification produces the value

$$(2.33) \qquad y_3 = A^3y_0 + B(1 + A + A^2).$$

The reader who sees (2.31), (2.32), and (2.33) should be willing to conjecture that the pattern visible in these expressions persists. *If* this is the case, then

$$(2.34) \quad y_k = A^k y_0 + B(1 + A + A^2 + \cdots + A^{k-1}) \qquad k = 1, 2, 3, \cdots.$$

We first simplify this expression and then prove that our conjecture is indeed a correct one.

The quantity in parenthesis in (2.34) is the sum of k terms of a geometric progression with first term 1 and common ratio A. Therefore (see 15 of Problems 1.6)

$$(2.35) \quad 1 + A + A^2 + \cdots + A^{k-1} = \begin{cases} \dfrac{1 - A^k}{1 - A} & \text{if } A \neq 1 \\ k & \text{if } A = 1 \end{cases}$$

and we may write

$$(2.36) \quad y_k = \begin{cases} A^k y_0 + B \dfrac{1 - A^k}{1 - A} & \text{if } A \neq 1 \\ y_0 + Bk & \text{if } A = 1. \end{cases} \qquad k = 0, 1, 2, \cdots$$

(Note that we include $k = 0$ since this formula does reduce to y_0 when $k = 0$.)

THEOREM 2.2. *The function y given by (2.36) is a solution, and the only solution, of the difference equation (2.30) with y_0 prescribed.*

PROOF. We have already noted that the solution (2.36) does in fact reduce to y_0 when $k = 0$. That we have a solution can be shown by direct substitution into the difference equation. From (2.36), with $A \neq 1$, we have

$$y_{k+1} = A^{k+1} y_0 + B \frac{1 - A^{k+1}}{1 - A},$$

so we must show that

$$(2.37) \quad A^{k+1} y_0 + B \frac{1 - A^{k+1}}{1 - A} = A \left(A^k y_0 + B \frac{1 - A^k}{1 - A} \right) + B$$

is an identity, i.e., true for $k = 0, 1, 2, \cdots$. It suffices to simplify the right-hand side. Now the right-hand side is equal to

$$A^{k+1} y_0 + B \left[1 + \frac{A(1 - A^k)}{1 - A} \right].$$

We have only to observe that

$$1 + \frac{A(1 - A^k)}{1 - A} = \frac{1 - A + A(1 - A^k)}{1 - A} = \frac{1 - A^{k+1}}{1 - A}$$

to complete the proof that in the case $A \neq 1$ we do indeed have a solution of the difference equation. If $A = 1$, a similar proof can be given (see 1 of Problems 2.4).

That this is the only solution for which y_0 is prescribed follows immediately from Theorem 2.1.

COROLLARY. *Let y be a solution of the difference equation* (2.30) *over the set $k = 0, 1, 2, \cdots$. Then there is a constant C for which*

$$(2.38) \quad y_k = \begin{cases} CA^k + B \dfrac{1 - A^k}{1 - A} & \text{if } A \neq 1 \\ C + Bk & \text{if } A = 1. \end{cases} \quad k = 0, 1, 2, \cdots$$

PROOF. We have only to equate the constant C with the value of the solution y at $k = 0$ and use Theorem 2.2.

Formula (2.38) gives *all* the solutions of the difference equation (2.30), one solution for each value of the arbitrary constant C. Remembering that solving a difference equation means finding all its solutions, we can say that we have now solved the difference equation (2.30).

Examples

(a) Suppose we are given the difference equation

$$y_{k+1} = 2y_k + 1 \quad k = 0, 1, 2, \cdots$$

with the initial condition $y_0 = 5$. From (2.36) we find (since $A = 2, B = 1$) the unique solution

$$y_k = 5 \cdot 2^k + 1 \cdot \frac{1 - 2^k}{1 - 2}$$

or

$$y_k = 6 \cdot 2^k - 1 \quad k = 0, 1, 2, \cdots.$$

If we write the consecutive values of y serially, starting with $y_0 = 5$, we have the sequence of increasing values 5, 11, 23, 47, 95, 191, $\cdots$.

(b) Given the difference equation

$$2y_{k+1} - y_k = 4 \quad k = 0, 1, 2, \cdots$$

with the initial condition $y_0 = 3$, we first rewrite the difference equation in the form (2.30):

$$y_{k+1} = \tfrac{1}{2}y_k + 2 \quad k = 0, 1, 2, \cdots.$$

Now, from (2.36) with $A = \tfrac{1}{2}$ and $B = 2$, we find the solution

$$y_k = (\tfrac{1}{2})^k \cdot 3 + 2 \frac{1 - (\tfrac{1}{2})^k}{1 - \tfrac{1}{2}}$$

which, upon simplification, may be written as

$$y_k = 4 - (\tfrac{1}{2})^k.$$

The sequence of values $y_0, y_1, y_2, \cdots$ is again an increasing sequence: $3, 3\frac{1}{2}, 3\frac{3}{4}, 3\frac{7}{8}, 3\frac{15}{16}, \cdots$. The reader should note an important difference between this increasing sequence and the one obtained in the preceding example. In the present case the values of y steadily increase but always remain below the value 4, whereas the y-values of Example (a) increase without bound.

(c) As a third example we consider the difference equation

$$y_{k+1} = -y_k + 1 \qquad k = 0, 1, 2, \cdots$$

with the initial condition $y_0 = 1$. We find from (2.36), with $A = -1$, $B = 1$,

$$y_k = (-1)^k + \frac{1 - (-1)^k}{1 - (-1)}$$

which, upon simplification, yields the solution

$$y_k = \tfrac{1}{2}[1 + (-1)^k] \qquad k = 0, 1, 2, \cdots.$$

If k is 0 or an even integer, then $(-1)^k = 1$; if k is an odd integer, then $(-1)^k = -1$. The quantity in brackets therefore alternates in value between 2 and 0. Thus we obtain the following sequence of y values: $1, 0, 1, 0, 1, 0, \cdots$. This solution may be described as showing oscillations of constant amplitude.

In the next sections we shall discuss in some detail the various possible behaviors of the sequence of y-values obtained from solving a linear first-order difference equation of the type given in (2.30). We conclude this section with some remarks about Theorem 2.2.

To determine a unique solution of the difference equation does not require the specification of the particular value y_0. Any other value of y will suffice. We have only to use whatever value of y is given to determine the constant C in the solution (2.38). For example, to find the solution of the difference equation

$$y_{k+1} = 2y_k - 1 \qquad k = 0, 1, 2, \cdots$$

for which $y_3 = 9$, we put $k = 3$ in (2.38) and find (with $A = 2$, $B = -1$)

$$y_3 = C \cdot 2^3 - \frac{1 - 2^3}{1 - 2} = 8C - 7.$$

Therefore, we choose $C = 2$ and have the solution

$$y_k = 2 \cdot 2^k - \frac{1 - 2^k}{1 - 2} \qquad \text{or} \qquad y_k = 2^k + 1 \qquad k = 0, 1, 2, \cdots.$$

Finally, we remark that the results already obtained do not depend upon the assumption, made up to now merely for convenience, that the difference equation is defined over the set of k-values $0, 1, 2, \cdots$. We can, in fact, prove the following result precisely as we proved Theorem 2.2. The proof is outlined in 4 of Problems 2.4.

THEOREM 2.3. *The linear first-order difference equation*

$$(2.39) \qquad y_{k+1} = Ay_k + B \qquad k = a, a+1, a+2, \cdots$$

taken over the indicated set of k-values (which may or may not continue indefinitely) has infinitely many solutions. If y is a solution, there is a constant C such that

$$(2.40) \quad y_k = \begin{cases} CA^{k-a} + B\,\dfrac{1 - A^{k-a}}{1 - A} & \text{if } A \neq 1 \\[2mm] C + B(k-a) & \text{if } A = 1. \end{cases} \qquad k = a, a+1, a+2, \cdots$$

If a single value of y is prescribed for one of the k-values a, $a+1$, $a+2$, $\cdots$, then a unique solution of (2.39) is determined. In particular, if y_a is prescribed, then the unique solution of (2.39) is given by (2.40) with $C = y_a$.

Note that when $a = 0$, these results reduce to those of Theorem 2.2 and its corollary.

PROBLEMS 2.4

1. Complete the proof of Theorem 2.2 by showing that the function (2.36) is a solution of a difference equation (2.30) when $A = 1$.

2. Each of the following difference equations is assumed to be defined over the set of k-values $0, 1, 2, \cdots$. In addition, suppose y_0 is prescribed in each case and is equal to 2. Find the solution of the difference equation, write out the first six values of y in sequence form, and describe the apparent behavior of y in this sequence.

   (a) $y_{k+1} = 3y_k - 1$
   (b) $y_{k+1} = y_k + 2$
   (c) $y_{k+1} + y_k - 2 = 0$
   (d) $3y_{k+1} = 2y_k + 3$
   (e) $2y_{k+1} + y_k - 3 = 0$
   (f) $y_{k+1} + 3y_k = 0$

3. Repeat Problem 2, assuming that the prescribed value of y is $y_0 = 6$. Again write out the first six values of y, starting with y_0. Does the behavior in this sequence change as the initial condition is changed from $y_0 = 2$ to $y_0 = 6$?

4. To prove Theorem 2.3 proceed as follows, starting with the difference equation (2.39): (a) Define the new index j by the relation $j = k - a$ so that the k-values $a, a+1, a+2, \cdots$ are transformed into j-values $0, 1, 2, \cdots$. (b) Note that

$$y_k = y_{a+j} = E^a y_j = z_j$$

where the last equality defines a new function z. Show that the difference equation (2.39) can be written in the form

$$z_{j+1} = Az_j + B \qquad j = 0, 1, 2, \cdots.$$

(c) Solve this difference equation using Theorem 2.2. [Steps (a) and (b) were designed to reduce the original problem to one which we have already solved.] (d) Obtain the results of Theorem 2.3 by replacing j by its value $k - a$.

5. Solve the difference equation

$$y_{k+1} = 2y_k + 3 \qquad k = 1, 2, 3, 4, 5.$$

Find the unique particular solution for which $y_1 = 0$. (Note that if y is a solution, then y must be defined for $k = 1, 2, \cdots, 6$.)

6. Write the solutions of the difference equations in Problem 2, assuming each to be defined over the set $k = 1, 2, 3, \cdots$. Assume that the prescribed value is $y_1 = 2$.

7. Often a nonlinear difference equation can be solved by reducing it to a corresponding linear difference equation. For example, consider the equation[2]

$$(2.41) \qquad f_{k+1} = \frac{f_k}{1 + f_k} \qquad k = 0, 1, 2, \cdots$$

with f_0 as a prescribed positive number.

(a) Show first that if $f_0 \neq 0$, then $f_k \neq 0$ for every k-value. This allows us to define a new function y, the reciprocal of f, by the equation

$$(2.42) \qquad y_k = \frac{1}{f_k} \qquad k = 0, 1, 2, \cdots.$$

(b) Show that the difference equation (2.41) is transformed into the linear difference equation

$$(2.43) \qquad y_{k+1} = y_k + 1 \qquad k = 0, 1, 2, \cdots.$$

(c) Solve (2.43) and thus show that $y_k = y_0 + k$. Then use (2.42) to find the required solution of the original difference equation:

$$(2.44) \qquad f_k = \frac{f_0}{1 + k f_0} \qquad k = 0, 1, 2, \cdots.$$

8. Using the same technique as that employed to obtain (2.36), show that the difference equation

$$(2.45) \qquad y_{k+1} = A y_k + B C^k \qquad k = 0, 1, 2, \cdots,$$

---

[2] In population genetics, (2.41) arises if f_k denotes the relative frequency of a recessive gene r in generation number k of a population undergoing random (or Mendelian) mating, except for pure recessive individuals (of genotype rr), who either do not survive or are sterile. Equation (2.44) shows the generation-by-generation reduction in proportion of the r gene under this eugenic selection force. See G. Dahlberg, *Mathematical Methods of Population Genetics*, Interscience Publishers, New York, 1948, p. 40; W. Feller, *An Introduction to Probability Theory and Its Applications*, Vol. I, 2nd ed., John Wiley, New York, 1957, p. 128; L. Hogben, *An Introduction to Mathematical Genetics*, W. W. Norton, New York, 1946, p. 127.

where A, B, and C are constants ($A \neq 0$), has the solution (with y_0 prescribed)

$$(2.46) \qquad y_k = \begin{cases} A^k y_0 + \dfrac{B(C^k - A^k)}{C - A} & \text{if } A \neq C \\ A^k y_0 + B A^{k-1} k & \text{if } A = C. \end{cases} \qquad k = 0, 1, 2, \cdots$$

Show that this equation and its solution reduce to (2.30) and (2.36) when $C = 1$.

2.5 SEQUENCES

We solve difference equations in order to obtain information about the function or functions which satisfy these equations. We have already seen, in the illustrative examples and problems of the preceding section, that solutions of even simple difference equations may show a wide variety of behaviors: some have values y_k which decrease as k increases; others have increasing values; still others oscillate between positive and negative values. The determination of conditions under which these and other behaviors are manifested is of fundamental importance in the applications of difference equations to problems of the social sciences.

For our present purposes, a solution y of the difference equation

$$y_{k+1} = A y_k + B \qquad k = 0, 1, 2, \cdots,$$

with y_0 prescribed, is most conveniently represented by enumerating its values in sequential order

$$(2.47) \qquad y_0, y_1, y_2, \cdots.$$

As a matter of fact, *any* function whose domain of definition is the set of consecutive integers $0, 1, 2, \cdots$ may be defined by such an enumeration since we are thereby given the value of the function at each number in its domain of definition. Because of this property, such a function is called a *sequence*. (Ordinarily a sequence is defined as a function whose domain of definition is the set of positive integers, $1, 2, 3, \cdots$. We use the equivalent set $0, 1, 2, \cdots$ for convenience in applying results about sequences to solutions of difference equations over this set of k-values. The important point is that the values of the function may be enumerated.) The enumeration in (2.47) is often denoted by $\{y_k\}$ and we speak of the sequence $\{y_k\}$, with *general term* y_k, whose elements or terms are the (real) numbers $y_0, y_1, y_2, \cdots$.

Since all the difference equations we shall consider are defined over sets of consecutive integers, their solutions are sequences. In particular, the solutions obtained in the preceding section are sequences. We shall best be able to learn about the behavior of these solutions by studying some of the elementary theory of sequences of real numbers.

The following are examples of sequences:

(a) $\{1\}$ $1, 1, 1, 1, 1, \cdots$

(b) $\{k - 1\}$ $-1, 0, 1, 2, 3, \cdots$

(c) $\{2^k\}$ $1, 2, 4, 8, 16, \cdots$

(d) $\left\{\dfrac{1}{k + 1}\right\}$ $1, \frac{1}{2}, \frac{1}{3}, \frac{1}{4}, \frac{1}{5}, \cdots$

(e) $\{1 - (\frac{1}{2})^k\}$ $0, \frac{1}{2}, \frac{3}{4}, \frac{7}{8}, \frac{15}{16}, \cdots$

(f) $\{(-\frac{1}{3})^k\}$ $1, -\frac{1}{3}, \frac{1}{9}, -\frac{1}{27}, \frac{1}{81}, \cdots$

(g) $\{(-2)^k\}$ $1, -2, 4, -8, 16, \cdots$

(h) $\{1 + (-1)^k\}$ $2, 0, 2, 0, 2, \cdots$

In studying the behavior of sequences of real numbers, several possibilities present themselves. It may happen that one can find a positive number M such that the inequality[3]

$$|y_k| \leq M \qquad \text{or} \qquad -M \leq y_k \leq M$$

is satisfied for every value of k. The sequence $\{y_k\}$ is then said to be *bounded*. If no such number M exists, the sequence is *unbounded*. Sequence (a) is easily seen to be bounded since for all k,

$$|1| = 1 \leq 1.$$

Similarly, we note that

$$\left|\frac{1}{k + 1}\right| = \frac{1}{k + 1} \leq 1 \qquad k = 0, 1, 2, \cdots$$

and

$$|1 + (-1)^k| \leq 2 \qquad k = 0, 1, 2, \cdots$$

so sequences (d) and (h) are bounded. Sequences (b), (c), and (g) are unbounded, but any number M greater than or equal to 1 will serve as a bound for sequences (e) and (f).

Let us look more closely at sequence (d) whose general term is $1/(k + 1)$. As we go out farther and farther in the sequence, the terms become smaller and smaller, getting nearer and nearer to the value 0. None of the terms is actually equal to 0, but if we go out far enough in sequence (d) we may be certain that each of its terms will differ from 0 by as little as we please. Sequence (f) is also of this type. Although its terms are alternately positive and negative, we may be certain, by going far enough out in the sequence (i.e., by taking k large enough), that its terms get (and stay) arbitrarily close to the number 0. We make these notions precise in the following definition.

---

[3] $|x|$ denotes the *absolute* (or numerical) value of the number x, i.e., $|x| = x$ if $x \geq 0$, but $|x| = -x$ if $x < 0$. For example, $|2| = 2$, $|-2| = 2$, $|-5| = 5$. If $|x| \leq M$, then $-M \leq x \leq M$ and conversely.

DEFINITION 2.5. *A sequence $\{y_k\}$ is called a null sequence if, corresponding to any positive number ε, no matter how small, there may be found a positive integer N (depending on ε), such that $|y_k| < \varepsilon$ for all $k \geq N$.*

The number ε may be thought of as a challenge number; we think of ourselves challenged to discover whether it is possible to go far enough out in the sequence (i.e., find a value N in response to the challenge number ε) in order to have all terms of the sequence from that point on less in absolute value than the challenge number ε. A null sequence is one for which this challenge may be successfully met for *every* positive challenge number, however small. Let us consider these ideas in the context of sequence (d) with general term $1/(k + 1)$. Suppose we are given the challenge number $\varepsilon = \frac{1}{3}$. Then we may respond with $N = 3$ since all terms from y_3 on are less than $\frac{1}{3}$, i.e.,

$$\left| \frac{1}{k + 1} \right| < \frac{1}{3} \quad \text{for all } k \geq 3.$$

If a different challenge number is presented to us, we will in general have to offer a different response value. This is what is meant by the parenthetic remark in Definition 2.5 that N depends on ε. Thus, if $\varepsilon = \frac{1}{10}$, we must go farther out in sequence (d) to get all terms less than $\frac{1}{10}$ than to get all terms less than $\frac{1}{3}$. In fact, we must go to the term $y_{10} = \frac{1}{11}$ before we meet the new challenge represented by $\varepsilon = \frac{1}{10}$, i.e.,

$$\left| \frac{1}{k + 1} \right| < \frac{1}{10} \quad \text{for all } k \geq 10.$$

We now prove that we can successfully meet *any* challenge number ε. If $\varepsilon \geq 1$ we have a trivial challenge since every term of the sequence after y_0 is less than ε. We may choose $N = 1$ to meet such a challenge number. Now let $0 < \varepsilon < 1$. We want to have

$$(2.48) \qquad \left| \frac{1}{k + 1} \right| = \frac{1}{k + 1} < \varepsilon.$$

This will be the case as soon as

$$k + 1 > \frac{1}{\varepsilon} \quad \text{or} \quad k > \frac{1}{\varepsilon} - 1.$$

We then respond to the challenge number ε by choosing for N any integer exceeding $(1/\varepsilon) - 1$. For all k greater than such an N we shall have $k > (1/\varepsilon) - 1$ and therefore also have (2.48) satisfied. Thus, we are able to satisfy the requirement in Definition 2.5 and this proves $\{1/(k + 1)\}$ is a null sequence.

Sequence (*e*), whose terms are

$$0, \tfrac{1}{2}, \tfrac{3}{4}, \tfrac{7}{8}, \tfrac{15}{16}, \cdots,$$

is not a null sequence, but it has one very important property in common with null sequences. It too has terms which seem to get nearer and nearer to some number as we go farther and farther out in the sequence. For null sequences this number is 0, whereas the terms of sequence (*e*) appear to be approaching the number 1. If we want to detect this property in a sequence, how should we proceed? Suppose we subtract 1 from every term of sequence (*e*) to form the new sequence

$$-1, -\tfrac{1}{2}, -\tfrac{1}{4}, -\tfrac{1}{8}, -\tfrac{1}{16}, \cdots$$

and suppose further that we are able to prove this new sequence is a null sequence. Our intuitive ideas about numbers would now support the following claim: the original sequence (*e*) must have terms which get closer and closer to 1 as we go farther and farther out in the sequence since the new sequence has terms 1 less than the corresponding terms in (*e*) and these new terms get close to 0.

These intuitive notions are formalized in the following very important definition.

DEFINITION 2.6. *If $\{y_k\}$ is a given sequence and there is a number L such that the sequence $\{y_k - L\}$ is a null sequence, then the given sequence is said to have the limit L (or converge to L). The elements y_k of the sequence are said to approach (or tend to or converge to) the limit L.*

The fact that a sequence $\{y_k\}$ has a limit L is expressed symbolically by writing

$$\lim_{k \to \infty} y_k = L$$

or more simply

$$y_k \to L \qquad \text{as } k \to \infty.$$

A sequence $\{y_k\}$ with a limit L is called *convergent*. A sequence without a limit is called *divergent*. It is clear that a null sequence is convergent with limit 0.

Let us use Definition 2.6 to prove that sequence (*e*) is convergent with a limit equal to 1. We must prove that after subtracting 1 from each term of sequence (*e*), the resulting sequence

$$\{1 - (\tfrac{1}{2})^k - 1\} = \{-(\tfrac{1}{2})^k\}$$

is a null sequence. If a challenge number $\varepsilon > 0$ is prescribed, we must be able to find a number N so that

$$(2.49) \qquad |-(\tfrac{1}{2})^k| < \varepsilon \qquad \text{whenever } k \geq N.$$

But

$$\left|-(\tfrac{1}{2})^k\right| = (\tfrac{1}{2})^k = \frac{1}{2^k} < \varepsilon$$

whenever

$$2^k > \frac{1}{\varepsilon},$$

so it suffices to choose for N any positive integer for which

(2.50) $$2^N > \frac{1}{\varepsilon}.$$

For example, if $\varepsilon = \tfrac{1}{10}$, then $1/\varepsilon = 10$. Since $2^4 > 10$ we may choose $N = 4$ and be certain that (2.49) is satisfied. If $\varepsilon = \tfrac{1}{100}$, then $1/\varepsilon = 100$. Since $2^7 > 100$ we may choose $N = 7$ and satisfy (2.49). That sequence (e) has the limit 1 follows from the fact that we can successfully respond to [i.e., satisfy (2.49) for] any challenge number ε by choosing N in accordance with (2.50).

Let us take a closer look at sequences (a), (d), (e), and (f), all of which are convergent, since they demonstrate a number of different possibilities. Sequence (a) has unchanging terms all equal to 1. The limit is 1 and all terms of the sequence are equal to this limiting value. We call such a sequence a *constant sequence*. Sequences (d), (e), and (f) are not of this variety. The terms of sequence (d) become steadily smaller and approach the limit 0 from above, the terms of sequence (e) become steadily larger and approach the limit 1 from below, and the terms of sequence (f) are alternately above and below the limiting value 0.

The following definition introduces the usual mathematical terminology for the description of such sequences.

DEFINITION 2.7. (a) *The sequence $\{y_k\}$ is said to be monotone (or steadily) increasing if $y_{k+1} > y_k$ for all k. Similarly, $\{y_k\}$ is monotone (or steadily) decreasing if $y_{k+1} < y_k$ for all k. (b) The sequence $\{y_k\}$, if convergent with limit L, is said to be damped oscillatory (around the value L) if each element of the sequence less than L is followed by some element (not necessarily the next one) greater than L, and vice versa.*

The reader should note that a damped oscillatory sequence is necessarily convergent according to Definition 2.7(b). But monotone sequences need not be convergent. This is clearly shown by sequence (e), with general term $y_k = 2^k$, which has no limit but is steadily increasing. It can be proved[4] that if a monotone sequence is bounded, then it is convergent. A monotone sequence which is unbounded must be one of two

[4] See R. Courant and H. Robbins, *What is Mathematics?* Oxford Univ. Press, New York, 1941, pp. 295–296.

types, typified by the sequences $\{2^k\}$ and $\{-2^k\}$. In the first the terms steadily increase, becoming more and more positive; the terms of the second decrease and become more and more negative as we go out in the sequence. To distinguish these cases we say that the former diverges to $+\infty$ (positive infinity) and the latter diverges to $-\infty$ (negative infinity). The precise meaning of these notions is embodied in the following definition.

DEFINITION 2.8. *The sequence $\{y_k\}$ is said to diverge to $+\infty$ (positive infinity) if, corresponding to any positive number P, however large, we can find a positive integer N (depending on P), such that $y_k > P$ for all $k \geq N$.*

If the sequence $\{y_k\}$ diverges to $+\infty$, we write $y_k \to +\infty$ as $k \to \infty$. A similar definition holds for sequences which diverge to $-\infty$. Note that sequences which diverge to $+\infty$ or to $-\infty$ need not be monotone (cf. 10 of Problems 2.5).

The reader's understanding of Definition 2.8 may be aided by thinking of the number P as a challenge number. To demonstrate that $y_k \to +\infty$ we must be able to meet this challenge by going far enough out in the sequence (i.e., finding a term y_N) so that all terms from that point on (i.e., y_k for all $k \geq N$) are larger than the challenge number P. Consider, for example, the sequence $\{2^k\}$ and let $P = 10$. Then we respond to this challenge by moving to the term 2^4 (i.e., $N = 4$) and observing that this, and all succeeding terms, exceed $P = 10$. If we are challenged with the larger number 100, then we may respond with $N = 7$, since all terms from 2^7 on are larger than 100. To say that the sequence $\{2^k\}$ diverges to $+\infty$ means that we can respond in this way to *any* positive challenge number. We need only respond to the challenge number P by choosing N so that $2^k > P$ for all $k \geq N$. The response $N = P$ (although not the smallest response number) certainly meets this requirement.

We conclude this survey of behaviors of sequences of real numbers by considering two additional ways by which a sequence may diverge. As prototypes we think of sequences (g) and (h) with terms $1, -2, 4, -8, 16, \cdots$ and $2, 0, 2, 0, 2, \cdots$ respectively. Both sequences diverge but neither diverges to $+\infty$ or to $-\infty$. The terms of the first are alternately positive and negative and increase in numerical value over all bounds, whereas the terms of the second oscillate between the finite values 2 and 0. We make the following definition to cover these behaviors.

DEFINITION 2.9. *If the sequence $\{y_k\}$ is divergent, but does not diverge to $+\infty$ or to $-\infty$, then $\{y_k\}$ is said to oscillate finitely if it is bounded, and to oscillate infinitely if it is unbounded.*

We may now summarize our findings: sequences are either convergent or divergent. We have defined four behaviors under each category; see

Table 2.1. (These are not the only possibilities, but they suffice for our purposes.)

Convergent Sequences	*Divergent Sequences*
C1. Constant	D1. Diverges to $+\infty$
C2. Bounded and monotone increasing	D2. Diverges to $-\infty$
C3. Bounded and monotone decreasing	D3. Oscillates finitely
C4. Damped oscillatory	D4. Oscillates infinitely

Some examples of each of the behaviors given in Table 2.1 are:

C1. Constant:

$\{2\}$ 2, 2, 2, 2, 2, $\cdots$ (limit = 2)

C2. Bounded, monotone increasing:

$\left\{\dfrac{3^k - 1}{3^k}\right\}$ $0, \frac{2}{3}, \frac{8}{9}, \frac{26}{27}, \frac{80}{81}, \cdots$ (limit = 1)

$\left\{5 - \dfrac{1}{2^k}\right\}$ $4, 4\frac{1}{2}, 4\frac{3}{4}, 4\frac{7}{8}, 4\frac{15}{16}, \cdots$ (limit = 5)

$\left\{\dfrac{2k}{k + 1}\right\}$ $0, 1, \frac{4}{3}, \frac{6}{4}, \frac{8}{5}, \cdots$ (limit = 2)

C3. Bounded, monotone decreasing:

$\left\{\dfrac{500}{k + 1}\right\}$ $\frac{500}{1}, \frac{500}{2}, \frac{500}{3}, \frac{500}{4}, \frac{500}{5}, \cdots$ (limit = 0)

$\left\{5 + \dfrac{1}{2^k}\right\}$ $6, 5\frac{1}{2}, 5\frac{1}{4}, 5\frac{1}{8}, 5\frac{1}{16}, \cdots$ (limit = 5)

C4. Damped oscillatory:

$\{1 + (-\frac{1}{2})^k\}$ $2, \frac{1}{2}, 1\frac{1}{4}, \frac{7}{8}, 1\frac{1}{16}, \cdots$ (limit = 1)

$\left\{\dfrac{1000}{(-5)^k}\right\}$ $1000, -200, 40, -8, 1\frac{3}{5}, \cdots$ (limit = 0)

D1. Diverges to $+\infty$:

$\{3^k\}$ 1, 3, 9, 27, 81, $\cdots$

$\{1.01^k\}$ 1, 1.01, 1.0201, 1.030301, $\cdots$

$\{k\}$ 0, 1, 2, 3, 4, $\cdots$

$\{k + (-1)^k\}$ 1, 0, 3, 2, 5, $\cdots$

D2. Diverges to $-\infty$:

$\{1 - 2^k\}$ $\qquad\qquad$ $0, -1, -3, -7, -15, \cdots$

$\{1 - 4k\}$ $\qquad\qquad$ $1, -3, -7, -11, -15, \cdots$

D3. Oscillates finitely:

$\{(-1)^k\}$ $\qquad\qquad$ $1, -1, 1, -1, 1, \cdots$

$\left\{5 + (-1)^k \dfrac{k}{k+1}\right\}$ $\qquad$ $5, 4\frac{1}{2}, 5\frac{2}{3}, 4\frac{1}{4}, 5\frac{4}{5}, \cdots$

D4. Oscillates infinitely:

$\{1 + (-2)^k\}$ $\qquad\qquad$ $2, -1, 5, -7, 17, \cdots$

$\{(-1)^k k\}$ $\qquad\qquad$ $0, -1, 2, -3, 4, \cdots$

PROBLEMS 2.5

1. Consider the following sequences $\{y_k\}$. In each case, write the first five terms of the sequence and determine its type by placing it in one of the classes C1–C4 or D1–D4.

(a) $y_k = \dfrac{1}{(k+2)^3}$ $\qquad\qquad$ (f) $y_k = \dfrac{k+1}{k+2}$

(b) $y_k = 10^k$ $\qquad\qquad\qquad$ (g) $y_k = 100 - 2^k$

(c) $y_k = \dfrac{1}{10^k}$ $\qquad\qquad\quad$ (h) $y_k = 2 + (-\frac{1}{2})^k$

(d) $y_k = \dfrac{k + (-1)^k}{k+1}$ $\qquad\quad$ (i) $y_k = \sin \dfrac{k\pi}{2}$

(e) $y_k = \dfrac{1}{100}(-5)^k$ $\qquad\quad$ (j) $y_k = \left(\dfrac{1}{k}\right)^{(-1)^k}$ $\quad (k \neq 0)$

2. Determine which of the sequences in Problem 1 are bounded and which unbounded. Is each bounded sequence convergent?

3. Use Definition 2.5 to prove that the sequence $\{1/10^k\}$ is a null sequence.

4. Use Definition 2.8 to prove that the sequence $\{10^k\}$ diverges to $+\infty$.

5. Use Definition 2.6 to prove that the sequence $\{(k + 1)/(k + 2)\}$ converges and has the limit 1.

6. Prove that if $\{y_k\}$ diverges to $+\infty$ and $z_k = -y_k$, then $\{z_k\}$ diverges to $-\infty$.

7. Consider the pairs of sequences (a) $\{(-1)^k\}$ and $\{(-1)^{k+1}\}$, (b) $\{(-1)^k\}$ and $\{(-1)^k\}$, and thus show that if two sequences, $\{y_k\}$ and $\{z_k\}$, both oscillate finitely, then their sum, the sequence $\{y_k + z_k\}$, may converge or oscillate finitely.

8. Consider the pairs of sequences (a) $\{k\}$ and $\{-k\}$, (b) $\{k^2\}$ and $\{-k\}$, (c) $\{k\}$ and $\{-k^2\}$, (d) $\{k + (-1)^k\}$ and $\{-k\}$, (e) $\{k^2 + (-1)^k k\}$ and $\{-k^2\}$, and thus show that if $\{y_k\}$ diverges to $+\infty$ and $\{z_k\}$ diverges to $-\infty$, then $\{y_k + z_k\}$ may converge, diverge to $+\infty$, diverge to $-\infty$, oscillate finitely, or oscillate infinitely.

9. Consider the pairs of sequences (a) $\{k\}$ and $\{2 + (-1)^k\}$, (b) $\{k\}$ and $\{-2 - (-1)^k\}$, (c) $\{k\}$ and $\{(-1)^k\}$, and thus show that if $\{y_k\}$ diverges to $+\infty$ and $\{z_k\}$ oscillates finitely, the product sequence $\{y_k z_k\}$ may diverge to $+\infty$, diverge to $-\infty$, or oscillate infinitely.

10. Show that the sequence $\{k + (-1)^k\}$ diverges to $+\infty$ but is not monotone increasing. Construct a sequence which diverges to $-\infty$ but is not monotone decreasing.

2.6 SOLUTIONS AS SEQUENCES

Our study of sequences may now be applied to the problem of determining the possible behaviors of solutions of the linear first-order difference equation

$$(2.51) \qquad y_{k+1} = Ay_k + B \qquad k = 0, 1, 2, \cdots.$$

We have found (Theorem 2.2) that the specification of the value y_0 determines the (unique) solution of this difference equation given by

$$(2.52) \qquad y_k = \begin{cases} A^k y_0 + B \dfrac{1 - A^k}{1 - A} & \text{if } A \neq 1 \\ y_0 + Bk & \text{if } A = 1. \end{cases} \qquad k = 0, 1, 2, \cdots$$

It is convenient to think of the prescribed initial value y_0 determining (or generating) the sequence $\{y_k\}$, with y_k defined by (2.52), the determination of the elements of the sequence being made in accordance with the difference equation (2.51). A definite sequence of real numbers is obtained as soon as we specify the particular initial value y_0 and the values of the constants A and B which determine the difference equation under consideration. Our aim in this section is to give a complete description of the possible behaviors of the solution sequences $\{y_k\}$ and, in particular, to determine the range of values of the three parameters y_0, A, and B for which each behavior type is exhibited.

The special case in which $A = 1$ is particularly simple and we dispose of it first.

THEOREM 2.4. *If $\{y_k\}$ is the solution (sequence) of the difference equation*

$$y_{k+1} = y_k + B \qquad k = 0, 1, 2, \cdots$$

with y_0 prescribed, then $\{y_k\}$ is a constant sequence if $B = 0$, diverges to $+\infty$ if $B > 0$, and diverges to $-\infty$ if $B < 0$.

PROOF. We know from (2.52) that $y_k = y_0 + Bk$. If $B = 0$, then $y_k = y_0$ for $k = 0, 1, 2, \cdots$ and $\{y_k\}$ is a constant sequence, as claimed. Now suppose $B > 0$. We must prove that $\{y_k\}$ diverges to $+\infty$, i.e., by Definition 2.8, given any positive number P, we must find a corresponding integer N so that $y_k > P$ for all $k \geq N$. Let the challenge number P be given. If $P \leq y_0$, then

$$y_k = y_0 + Bk > P$$

for all k greater than 0, so we may take $N = 1$ and be sure that $y_k > P$ for all $k \geq N$. If $P > y_0$, then we first solve the required inequality

$$y_k = y_0 + Bk > P$$

for k. We find (remembering that if $x > y$, $ax > ay$ if $a > 0$, but $ax < ay$ if $a < 0$)

$$Bk > P - y_0 \qquad k > \frac{P - y_0}{B}.$$

It is now easy to see that if we choose for N any integer greater than $(P - y_0)/B$ we shall have $y_k > P$ for all $k \geq N$. A similar argument for the case $B < 0$ completes the proof of the theorem.

We turn now to the difference equation (2.51) in the case $A \neq 1$ and first rewrite its solution in a more convenient form. From (2.52)

$$y_k = A^k y_0 + \frac{B}{1 - A} - \frac{B}{1 - A} A^k$$

$$= A^k \left(y_0 - \frac{B}{1 - A} \right) + \frac{B}{1 - A}$$

so that

$$y_k - \frac{B}{1 - A} = A^k \left(y_0 - \frac{B}{1 - A} \right)$$

or

(2.53) $$y_k - y^* = A^k(y_0 - y^*) \qquad k = 0, 1, 2, \cdots$$

if we simplify our notation by writing

(2.54) $$y^* = \frac{B}{1 - A}.$$

From (2.53) it is clear that if the prescribed initial value y_0 is equal to the number y^*, then no matter what the value of A,

$$y_k - y^* = 0 \qquad \text{or} \qquad y_k = y^*$$

for all values of k. But if y_0 is not equal to y^*, then y_k depends on A^k and the behavior of the sequence $\{y_k\}$ will depend upon the behavior of the sequence $\{A^k\}$. The following theorem summarizes the results we need for this most important special sequence.

THEOREM 2.5.   *The sequence $\{A^k\}$ is convergent if $-1 < A \leq 1$, divergent otherwise.   If $A = 0$ or $A = 1$, the sequence is constant (with limit 0 and 1 respectively); if $0 < A < 1$, it is monotone decreasing with limit 0; if $-1 < A < 0$, it is damped oscillatory with limit 0; if $A = -1$, it oscillates finitely; if $A < -1$, it oscillates infinitely; if $A > 1$, it diverges to $+\infty$.*

We shall need to know a useful inequality in order to complete the proof of this theorem.

LEMMA.   If $x > -1$, then

$$(1 + x)^k \geq 1 + kx \qquad k = 0, 1, 2, \cdots.$$

This lemma is easily proved by mathematical induction (cf. 4 of Problems 2.6).

We now proceed to prove Theorem 2.5. If $A = 0$ or $A = 1$, the result is obvious.[5] If $A = -1$, the indicated result is also immediate since the sequence $\{A^k\}$ becomes $1, -1, 1, -1, 1, -1, \cdots$.

Now consider the case $A > 1$. Then we may write $A = 1 + x$ with $x > 0$ and apply the lemma to obtain

$$A^k = (1 + x)^k \geq 1 + kx > kx.$$

Thus, if any positive number P is given, then $A^k > P$ for all $k \geq P/x$. We let N be any integer greater than P/x and then know that $A^k > P$ for all $k \geq N$. This proves that the sequence $\{A^k\}$ diverges to $+\infty$ if $A > 1$.

If $A < -1$, then the terms of the sequence $\{A^k\}$ are alternately positive and negative and so, although divergent, the sequence neither diverges to $+\infty$ nor to $-\infty$. The reasoning of the preceding paragraph may be used to show that the positive terms of the sequence increase without bound. The entire sequence is then certainly unbounded and therefore oscillates infinitely.

For the moment we combine the ranges $0 < A < 1$ and $-1 < A < 0$ and prove that $\{A^k\}$ converges to 0 (i.e., is a null sequence) if $-1 < A < 1$ If A is in this range and $A \neq 0$, we can write

$$|A| = \frac{1}{1 + x}$$

where x is some positive number.   Hence, by the lemma,

$$\frac{1}{|A|^k} = (1 + x)^k \geq 1 + kx > kx$$

---

[5] A very interesting commentary on the use of the word "obvious" in mathematics may be found in G. H. Hardy, *A Course of Pure Mathematics*, 9th ed., Macmillan, New York, 1949, footnote on p. 130.

or

$$0 < |A^k| = |A|^k < \frac{1}{kx}.$$

To proceed in accordance with Definition 2.5, with $\varepsilon > 0$ given, we let N be any integer greater than $1/(\varepsilon x)$. Then [since $1/(kx) < \varepsilon$ if $k > 1/(\varepsilon x)$],

$$|A|^k < \varepsilon \qquad \text{for all } k \geq N$$

so that $\{A^k\}$ is a null sequence.

If $0 < A < 1$, we have in addition to prove that $\{A^k\}$ is monotone decreasing. This is obvious since $A^{k+1} = A \cdot A^k < A^k$ in this case.

If $-1 < A < 0$, we must prove that $\{A^k\}$ is damped oscillatory. Since we have already shown that the sequence is convergent with limit 0, it suffices to observe that the terms of the sequence are alternately positive and negative. The proof of Theorem 2.5 is now complete. (Cf. 6 and 7 of Problems 2.6.)

Theorem 2.5 tells us about the behavior of A^k as k gets large, but we must study the behavior of y_k, which from (2.53) is given by

(2.55) $$y_k = A^k(y_0 - y^*) + y^*.$$

To find y_k we must first multiply A^k by the constant $(y_0 - y^*)$ and then add the constant y^* to the result. To study the sequence $\{y_k\}$ therefore requires that we discover how the behavior of a sequence is affected by first multiplying each term by a constant and then adding some constant to each term of the resulting sequence. For example, the bounded monotone increasing sequence $\{k/(k + 1)\}$ has limit 1. If we multiply each term by the constant 2, then the resulting sequence $\{(2k)/(k + 1)\}$ is still bounded and monotone increasing, but now has limit 2. However, if we multiply each term by the constant -2, then the new sequence $\{(-2k)/(k + 1)\}$ is bounded and monotone *decreasing* with limit -2. And if we multiply by the constant 0 we obtain a constant sequence all of whose terms are 0. In this same way, the reader may explore for himself the outcomes of multiplying (and adding) constants to each term of various types of sequences. By the *type* of sequence we mean its limiting behavior, i.e., type C1 (constant), C2 (bounded, monotone, increasing), $\cdots$, type D4 (oscillates infinitely), as listed in Table 2.1.

THEOREM 2.6. *Let $\{y_k\}$ be any sequence and c any constant. If $\{y_k\}$ converges with limit L, then $\{cy_k\}$ and $\{y_k + c\}$ also converge with limits cL and $L + c$ respectively. If $\{y_k\}$ diverges, then $\{cy_k\}$ diverges (if $c \neq 0$) and $\{y_k + c\}$ also diverges. Moreover, the sequences $\{cy_k\}$ and $\{y_k + c\}$ are both of the same type as $\{y_k\}$ with the following exceptions: (a) for any $\{y_k\}$, if $c = 0$, then $\{cy_k\}$ is a constant sequence (with all terms equal to 0);*

(b) *if* $c < 0$, *then* $\{cy_k\}$ *is of type* C3 *if* $\{y_k\}$ *is of type* C2 (*and vice versa*) *and* $\{cy_k\}$ *is of type* D1 *if* $\{y_k\}$ *is of type* D2 (*and vice versa*).

PROOF. We shall prove in detail that if $y_k \to L$ as $k \to \infty$, then $cy_k \to cL$ and $y_k + c \to L + c$. The remainder of the proof will be left as an exercise for the reader.

Let $\varepsilon > 0$ be given. In view of Definitions 2.5 and 2.6 we must show that positive integers N_1 and N_2 exist so that

$$(2.56) \quad |cy_k - cL| = |c(y_k - L)| = |c| \cdot |y_k - L| < \varepsilon \qquad \text{for } k \geq N_1$$

and

$$(2.57) \quad |(y_k + c) - (L + c)| = |y_k - L| < \varepsilon \qquad \text{for } k \geq N_2.$$

By hypothesis, $y_k \to L$ as $k \to \infty$ so if the challenge number ε is offered, we know the challenge is met by the existence of some positive integer, say N_3, for which

$$(2.58) \quad |y_k - L| < \varepsilon \qquad \text{if } k \geq N_3.$$

But this challenge can be met for *any* challenge number. Let N_4 denote the response for challenge number $\varepsilon/|c|$ if $c \neq 0$, i.e.,

$$(2.59) \quad |y_k - L| < \frac{\varepsilon}{|c|} \qquad \text{if } k \geq N_4.$$

Requirement (2.56) is now easily met by choosing $N_1 = 1$ if $c = 0$ and $N_1 = N_4$ if $c \neq 0$; (2.57) is satisfied by choosing $N_2 = N_3$. The existence of N_1 and N_2 proves that $cy_k \to cL$ and $y_k + c \to L + c$ as $k \to \infty$.

The results of this theorem, together with those of Theorem 2.5, allow us to determine the behavior of the (solution) sequence

$$(2.60) \quad \{y_k\} = \{A^k(y_0 - y^*) + y^*\}$$

for all possible combinations of values of y_0, A, and B. For example, if we put $A = B = \frac{1}{2}$ in (2.51), our difference equation becomes

$$(2.61) \quad y_{k+1} = \tfrac{1}{2}y_k + \tfrac{1}{2} \qquad k = 0, 1, 2, \cdots$$

and the solution, given by (2.55), is

$$y_k = (\tfrac{1}{2})^k(y_0 - 1) + 1.$$

In this case

$$y^* = \frac{B}{1 - A} = \frac{\frac{1}{2}}{1 - \frac{1}{2}} = 1.$$

Now Theorem 2.5 tells us $\{(\tfrac{1}{2})^k\}$ is convergent with limit 0 and is of type C3 (bounded and monotone decreasing). If $y_0 > 1$, then we are multiplying $\{(\tfrac{1}{2})^k\}$ by the positive constant $(y_0 - 1)$ so the resulting sequence $\{(\tfrac{1}{2})^k(y_0 - 1)\}$ is again a monotone decreasing null sequence by Theorem 2.6 [with $L = 0$ and $c = (y_0 - 1) > 0$]. Finally, we obtain the sequence

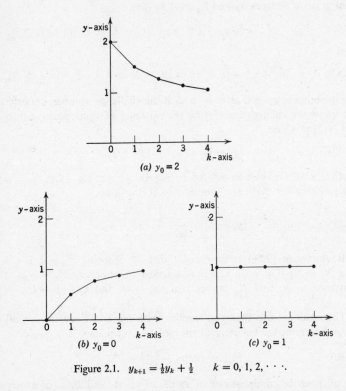

Figure 2.1. $y_{k+1} = \tfrac{1}{2}y_k + \tfrac{1}{2}$ $k = 0, 1, 2, \cdots$.

$\{y_k\}$ by adding the constant 1 and again by Theorem 2.6 we deduce that the solution sequence

$$\{y_k\} = \{(\tfrac{1}{2})^k(y_0 - 1) + 1\}$$

is a monotone decreasing convergent sequence with limit 1, if $y_0 > 1$. A similar argument shows that this solution is convergent with limit 1, but monotone increasing, if $y_0 < 1$. And, of course, if $y_0 = 1$, then $\{y_k\}$ is again convergent with each term equal to the limiting value 1.

These three possible behaviors for the solution of the difference equation (2.61) may conveniently be indicated graphically. In Figure 2.1(a), we have used the special value $y_0 = 2$ as typical of values greater than 1. Along the horizontal axis the values of k appear in sequential order starting with $k = 0$, and the values of y are measured along the vertical axis. Each point thus represents the pair of values (k, y_k), the point being placed on the vertical line through the value k on the horizontal axis at a height y_k above this axis. (If y_k is negative, then we move below the k-axis rather than above.) The y-values, starting with $y_0 = 2$, are 2, $1\frac{1}{2}$, $1\frac{1}{4}$, $1\frac{1}{8}$, $\cdots$. The resulting points are joined by straight-line segments in order to aid in seeing the way in which these values vary as k

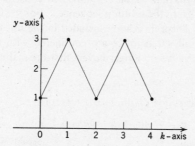

Figure 2.2. $y_{k+1} = -y_k + 4$, $y_0 = 1$.

increases. In Figure 2.1(b) we have used the special value $y_0 = 0$, and in Figure 2.1(c), the value $y_0 = 1$. Note how the graphs clearly exhibit the monotone decreasing, monotone increasing, and constant behaviors of the sequences $\{y_k\}$, in each case convergent with limit 1.

Let us consider another example. If $A = -1$ and $B = 4$, then the difference equation under consideration is

$$(2.62) \qquad y_{k+1} = -y_k + 4 \qquad k = 0, 1, 2, \cdots$$

and (2.55) yields the solution

$$(2.63) \qquad y_k = (-1)^k(y_0 - 2) + 2$$

since

$$y^* = \frac{B}{1 - A} = \frac{4}{1 - (-1)} = 2.$$

Starting with $\{(-1)^k\}$ which oscillates finitely, we first multiply by $(y_0 - 2)$ and then add 2 and Theorem 2.6 tells us that we end up with a sequence of the same type as the one with which we began unless $y_0 = 2$. Thus the solution of the difference equation (2.62) oscillates finitely unless

$y_0 = 2$. In Figure 2.2 we have graphed this solution using the value $y_0 = 1$ (in which case the sequence of y-values is 1, 3, 1, 3, 1, 3, $\cdots$).

Table 2.2 summarizes the behavior of the solution sequence

$$\{y_k\} = \{A^k(y_0 - y^*) + y^*\}$$

for all difference equations of the form

$$y_{k+1} = Ay_k + B \qquad k = 0, 1, 2, \cdots$$

(i.e., for all values of the constants A and B) and for all possible initial values y_0. The results in this table are immediate consequences of applying Theorems 2.5 and 2.6 to the solution sequence. For the sake of completeness the results of Theorem 2.4 are included in the last three lines of the table. Note that all values of A are considered with the exception of $A = 0$, in which case we no longer have a first-order difference equation.

TABLE 2.2

BEHAVIOR OF THE SOLUTION SEQUENCE $\{y_k\}$

$$y_{k+1} = Ay_k + B \qquad k = 0, 1, 2, \cdots$$

	Hypotheses			Conclusions	
Row	A	B	y_0	for $k = 1, 2, 3, \cdots$	the sequence $\{y_k\}$ is
(a)	$A \neq 1$		$y_0 = y^*$	$y_k = y^*$	constant $(= y^*)$
(b)	$A > 1$		$y_0 > y^*$	$y_k > y^*$	monotone increasing, diverges to $+\infty$
(c)	$A > 1$		$y_0 < y^*$	$y_k < y^*$	monotone decreasing, diverges to $-\infty$
(d)	$0 < A < 1$		$y_0 > y^*$	$y_k > y^*$	monotone decreasing, converges to limit y^*
(e)	$0 < A < 1$		$y_0 < y^*$	$y_k < y^*$	monotone increasing, converges to limit y^*
(f)	$-1 < A < 0$		$y_0 \neq y^*$		damped oscillatory, converges to limit y^*
(g)	$A = -1$		$y_0 \neq y^*$		divergent, oscillates finitely
(h)	$A < -1$		$y_0 \neq y^*$		divergent, oscillates infinitely
(i)	$A = 1$	$B = 0$		$y_k = y_0$	constant $(= y_0)$
(j)	$A = 1$	$B > 0$		$y_k > y_0$	monotone increasing, diverges to $+\infty$
(k)	$A = 1$	$B < 0$		$y < y_0$	monotone decreasing, diverges to $-\infty$

The various behavior types listed in Table 2.2 are roughly sketched in Figure 2.3. In each case, the graph is selected to be typical of the type of solution obtained for the specified values of A, B, and y_0. The graphs are labeled to match the rows of Table 2.2.

In the sections to follow we shall apply these results to a wide variety of difference equations which arise in the social and behavioral sciences. For purposes of easy reference, we summarize our findings in the following theorem.

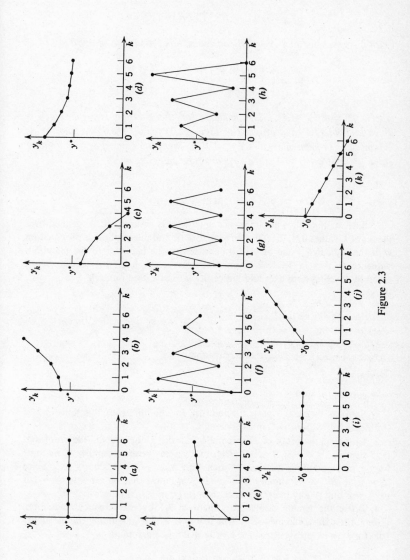

Figure 2.3

THEOREM 2.7. *The linear first-order difference equation*

$$y_{k+1} = Ay_k + B \qquad k = 0, 1, 2, \cdots$$

(with y_0 prescribed and A and B constants) has the unique solution

$$y_k = \begin{cases} A^k(y_0 - y^*) + y^* & \text{if } A \neq 1 \\ y_0 + Bk & \text{if } A = 1 \end{cases}$$

where $y^* = B/(1 - A)$. The behavior of the solution sequence $\{y_k\}$ is given in Table 2.2 (and graphed in Figure 2.3) for all possible values of A, B, and y_0. In particular, if $-1 < A < 1$, then $\{y_k\}$ converges with limiting value y^*, and if $|A| > 1$, then $\{y_k\}$ diverges unless $y_0 = y^*$.

PROBLEMS 2.6

1. Each of the following difference equations is assumed to be defined over the set of k-values $0, 1, 2, \cdots$. In each case (i) find the solution of the equation with the indicated value of y_0, (ii) characterize the behavior of the (solution) sequence $\{y_k\}$ as in Table 2.2, and (iii) draw a graph of this sequence $\{y_k\}$, carefully labeling both axes and indicating the scale used on each.

(a) $y_{k+1} = 3y_k - 1$ $\qquad y_0 = \frac{1}{2}$

(b) $y_{k+1} = 3y_k - 1$ $\qquad y_0 = 1$

(c) $y_{k+1} + 3y_k + 1 = 0$ $\qquad y_0 = 1$

(d) $2y_{k+1} - y_k = 2$ $\qquad y_0 = 4$

(e) $3y_{k+1} + 2y_k = 1$ $\qquad y_0 = 1$

(f) $y_{k+1} = y_k - 1$ $\qquad y_0 = 5$

(g) $y_{k+1} + y_k = 0$ $\qquad y_0 = -1$

2. Same as Problem 1 parts (ii) and (iii), for the difference equations in 2 of Problems 2.4, with $y_0 = 2$ in each case.

3. Complete the proof of Theorem 2.4 by writing out in detail the proof that $\{y_k\}$ diverges to $-\infty$ if $B < 0$. [*Hint:* Given any positive number P, we must find a corresponding number N such that $y_k < -P$ for all $k \geq N$. Since $y_k = y_0 + Bk < -P$ implies $Bk < -P - y_0$, show that we can take $N = 1$ if $y_0 \leq -P$ and N any integer greater than $[-(P + y_0)]/B$ otherwise.]

4. Prove the lemma needed for Theorem 2.5 by mathematical induction. [*Hint:* First show the inequality true if $k = 0$. Then, assuming the inequality true for $k = m$, prove it true for $k = m + 1$; i.e., assuming

(2.64) $$(1 + x)^m \geq 1 + mx,$$

prove that

$$(1 + x)^{m+1} \geq 1 + (m + 1)x.$$

Multiply both sides of (2.64) by the quantity $(1 + x)$, which is positive (why?), to get

$$(1 + x)^{m+1} \geq (1 + mx)(1 + x).$$

Complete the proof by showing that

$$(1 + mx)(1 + x) \geq 1 + (m + 1)x.]$$

5. Prove that if $\{y_k\}$ is convergent with limit L, then the sequence $\{y_{k+1}\}$ also converges to L and if $\{y_k\}$ diverges to $+\infty$ or $-\infty$, then $\{y_{k+1}\}$ does likewise.[6] (Illustration: If $\{y_k\}$ is the sequence $0, 1, 2, 3, \cdots$ with general term k, then $\{y_{k+1}\}$ is the sequence $1, 2, 3, 4, \cdots$ with general term $k + 1$; if $\{y_k\}$ is the sequence $\{1/(k + 1)\}$: $1, \frac{1}{2}, \frac{1}{3}, \cdots$, then $\{y_{k+1}\}$ is the sequence $\{1/(k + 2)\}$: $\frac{1}{2}, \frac{1}{3}, \frac{1}{4}, \cdots$. In general, $\{y_{k+1}\}$ is the sequence $\{y_k\}$ with first term omitted.)

6. If $A > 1$, show that $\{A^k\}$ is monotone increasing and always greater than 1. Then conclude that $\{A^k\}$ either converges to some limit L (which must be greater than 1) or diverges to $+\infty$. To prove that the latter alternative is correct, suppose that $\{A_k\}$ actually converges to L and deduce a contradiction as follows: use Problem 5 to show that

$$L = \lim_{k \to \infty} A^{k+1} = \lim_{k \to \infty} A \cdot A^k = A \lim_{k \to \infty} A^k = AL.$$

But it is impossible that $L = AL$ since both A and L are greater than 1. Hence $\{A^k\}$ diverges to $+\infty$ if $A > 1$.

7. If $0 < A < 1$, show that $\{A_k\}$ is monotone decreasing and must therefore converge to some limit, say L, or diverge to $-\infty$. Reject the second possibility by noting that A^k is always positive. Using an argument similar to that of Problem 6, show that $L = AL$ and thus prove that $L = 0$.

8. Complete the proof of Theorem 2.6 by showing that the sequences $\{cy_k\}$ and $\{y_k + c\}$ are both of the same type as $\{y_k\}$ with the exceptions noted. Consider each of the types C1, $\cdots$, C4, D1, $\cdots$, D4 separately.

2.7 SIMPLE AND COMPOUND INTEREST

If a sum of money earns *simple interest* at the rate r (i.e., percentage annual rate $100r\%$), the amount on deposit at any interest date is equal to the amount on deposit 1 year before that date plus the interest earned in that year on the *initial* principal invested. If we denote by S_k the sum on deposit after year number k and let S_0 equal the initial deposit, we can express this simple interest law symbolically in the form

$$(2.65) \qquad S_{k+1} = S_k + rS_0 \qquad k = 0, 1, 2, \cdots.$$

---

[6] After thinking through this result the reader should convince himself that limit properties of a sequence are unchanged by the omission of *any finite number* of terms from the sequence.

This is a linear first-order difference equation of the type considered in the preceding section. We find its unique solution (with S_0 prescribed) by invoking[7] Theorem 2.7 with $A = 1$ and $B = rS_0$. We find

$$S_k = S_0 + rS_0 k$$

or

(2.66) $$S_k = S_0(1 + kr) \qquad k = 0, 1, 2, \cdots.$$

This is the well-known *simple interest formula*, expressing the amount after k years as a function of the initial amount, the number of years at which this amount has been accumulating, and the annual simple interest rate. Of course, if $r > 0$ and $S_0 > 0$, then $B > 0$ and (see Table 2.2) the sequence $\{S_k\}$ diverges to $+\infty$.

The situation is different when the initial amount accumulates at *compound interest*. We let i stand for the interest rate per conversion period (e.g., if the annual rate is 4%, compounded semiannually, then $i = 0.02$; the same annual rate compounded quarterly makes $i = 0.01$, etc.). Again let S_k denote the sum on deposit at the end of k conversion periods and S_0 the initial sum deposited. The interest earned in any period is now computed on the total sum on deposit at the beginning of that period rather than on the initial principal, as in simple interest. Thus we obtain the equation

$$S_{k+1} = S_k + iS_k \qquad \text{or} \qquad S_{k+1} = (1 + i)S_k \qquad k = 0, 1, 2, \cdots$$

which is again a linear first-order difference equation with constant coefficients. From Theorem 2.7, with $A = (1 + i)$ and $B = 0$, we obtain the solution

(2.67) $$S_k = (1 + i)^k S_0$$

which is the standard compound interest formula. Table 2.2 shows that here, as in the case of simple interest, the sum on deposit diverges to $+\infty$ (since $1 + i > 1$). Note, however, from the typical graphs (b) and (j) in Figure 2.3, that the growth of S in simple interest is linear whereas it is more rapidly increasing in the case of compound interest.

Amortization is a method of repaying a debt, including both principal and interest, by a series of periodic payments, usually equal in amount, each of which is part payment of interest and part payment to reduce

[7] In Theorem 2.7 (and throughout the preceding sections) we have considered the function y and yet we are applying these results to the difference equation (2.65) for the function S. It should be clearly understood that Theorem 2.7 tells us about *any* function whose values are related as indicated therein. Equation (2.65) says that S is such a function. The *name* of the function is irrelevant.

outstanding principal. Suppose that the debt to be repaid is A, interest charges are at the compound rate i per conversion period, and the periodic payment (made at the end of each conversion period) is R. Let P_k be the outstanding principal after the kth payment of R (or, equivalently, at the beginning of the $(k + 1)$st period). Before the next payment is made, the debt increases by the interest due on the principal P_k. After the $(k + 1)$st payment of R, the outstanding debt is P_{k+1}. Hence

$$P_{k+1} = P_k + iP_k - R$$

or

(2.68) $$P_{k+1} = (1 + i)P_k - R \qquad k = 0, 1, 2, \cdots.$$

In addition, we know the initial debt, i.e.,

(2.69) $$P_0 = A.$$

The solution of this difference equation with the prescribed initial condition can be found by Theorem 2.7. We must identify the A, B, and y_0 of that theorem with $(1 + i)$, $-R$, and $P_0 = A$ respectively. Then

$$P^* = \frac{-R}{1 - (1 + i)} = \frac{R}{i}$$

and

$$P_k = (1 + i)^k \left(A - \frac{R}{i} \right) + \frac{R}{i}$$

or

(2.70) $$P_k = A(1 + i)^k - R \frac{(1 + i)^k - 1}{i} \qquad k = 0, 1, 2, \cdots.$$

The first term on the right is, by (2.67), the amount to which the initial debt accumulates (at the compound interest rate i) after k periods; the second term is the amount to which the k periodic payments accumulate in this same time (cf. 3 of Problems 2.7). Their difference is the remaining debt.

Suppose now that we wish to determine the periodic payment R in order to amortize the debt A by exactly n payments. Then $P_n = 0$ in (2.70) so that R may be determined from the equation

$$0 = A(1 + i)^n - R \frac{(1 + i)^n - 1}{i}.$$

We find

$$R = A \frac{i}{1 - (1 + i)^{-n}}$$

or

$$(2.71) \qquad R = A \frac{1}{a_{\overline{n}|i}}$$

if we define

$$(2.72) \qquad a_{\overline{n}|i} = \frac{1 - (1 + i)^{-n}}{i}.$$

The number $a_{\overline{n}|i}$, referred to as the *amortization factor*, is the present value of an annuity of 1 per period for n periods at the interest rate i (cf. 5 of Problems 2.7). The numbers $a_{\overline{n}|i}$ and $1/a_{\overline{n}|i}$ are tabulated[8] for the common values of n and i. For example, we find $1/a_{\overline{5}|.05} = 0.23097$ and $1/a_{\overline{10}|.05} = 0.12950$, so that five annual payments of \$23.10, or ten annual payments of \$12.95, will repay a debt of \$100 with interest at 5% compounded annually. The schedule shown in Table 2.3 exhibits the manner in which the five payments of \$23.10 successively lower the outstanding principal.

TABLE 2.3

AMORTIZATION SCHEDULE

Year	Outstanding Principal at Beginning of Year	Annual Payment at End of Year	Annual Payment is Divided into	
			Interest at 5% due at End of Year	Repayment of Principal at End of Year
1	\$100.00	\$23.10	\$5.00	\$18.10
2	81.90	23.10	4.10	19.00
3	62.90	23.10	3.14	19.96
4	42.94	23.10	2.15	20.95
5	21.99	23.10	1.10	22.00
6	0			
Totals		\$115.50	\$15.49	\$100.01

This schedule was constructed one row at a time. Thus at the end of year 1, the \$100 principal earns \$5.00 interest due; the \$23.10 payment is used to pay this \$5.00 with \$18.10 remaining to reduce the indebtedness to $100.00 - 18.10$ or \$81.90, etc. Note that as the amortization proceeds, the principals repaid increase and the interest charges decrease. (The discrepancy between the \$100.01 of principal repaid and the original \$100.00 principal is due to the fact that the annual payment is actually \$23.09748.)

---

[8] W. L. Hart, *Tables for Mathematics of Investment*, 4th ed., D. C. Heath, Boston, 1958, or any textbook on the mathematics of finance.

If a given debt is to be amortized and we are given a periodic payment, then an amortization schedule may be constructed. This schedule will show the total number of full payments to be made and will, in general, indicate the necessity for some additional final partial payment (cf. 7 of Problems 2.7). The retirement of a debt by the accumulation of a sinking fund is considered in 8–12 of Problems 2.7.

PROBLEMS 2.7

1. Find the number of years required for a given sum of money to double itself at (a) simple interest rate 2% per year, (b) compound interest rate 2% per year. [*Hint:* Put $S_k = 2S_0$ in (2.66) and (2.67) and solve for k. You will need to know the value of k for which $(1.02)^k = 2$. Tables of $(1 + i)^k$ are available for common values of i and k. From these one finds $(1.02)^{35} = 1.9998896$.]

2. Show that if the amount $A(1 + i)^{-k}$ is left to accumulate interest at the compound rate i, the resulting sum after k periods is A. [The quantity $A(1 + i)^{-k}$ is said to be the *discounted* value of A, discounted for k compound interest periods.]

3. Suppose the constant sum R is deposited at the end of each conversion period in a bank which credits interest at the compound rate i per period. Let A_k denote the total amount in the account at the end of k conversion periods. Show that

$$A_{k+1} = (1 + i)A_k + R \qquad k = 0, 1, 2, \cdots$$

with $A_0 = 0$. Solve this difference equation and thus show that

(2.73) $$A_k = Rs_{\overline{k}|i}$$

where

$$s_{\overline{k}|i} = \frac{(1 + i)^k - 1}{i}.$$

[An *ordinary annuity* is a set of periodic payments, usually equal in amount, payable at equal intervals of time, with payments being made at the end of each payment interval. The *amount* (or final value) of an ordinary annuity is defined as the sum of the amounts of all the payments, accumulated at compound interest to the time of the last payment. Formula (2.73) gives the amount of an ordinary annuity of k payments of R at the compound rate i per period under the assumption that the payment interval equals the conversion period.]

4. Using the value $s_{\overline{20}|.03} = 26.8704$ and the result of Problem 3, find (a) the amount of an ordinary annuity at 3% after 20 yearly payments of $100, (b) the yearly payment required in order to have 20 payments amount to $5000 at 3%.

5. The *term* of an annuity is defined as the time from the beginning of the first payment interval to the end of the last one. The *present value* of an annuity is its value at the beginning of the term. Prove that

$$s_{\overline{n}|i} = (1 + i)^n a_{\overline{n}|i}$$

and thus show that $a_{\overline{n}|i}$ is the present value of an annuity of 1 per period for n periods at the interest rate i.

6. Since $1/a_{\overline{10}|.05} = .0.12950$, ten annual payments of \$12.95 will amortize a debt of \$100 with interest at 5% compounded annually. Construct an amortization schedule for the extinction of this debt.

7. A debt of \$100 is to be amortized by equal payments of \$30 at the end of each year, plus a final partial payment 1 year after the last \$30 is paid. If interest is at the rate 5% compounded annually, construct an amortization schedule and thus show that three full \$30 payments are required together with a final partial payment of \$22.25 at the end of the fourth year.

8. In the *sinking fund* method of repaying a debt, it is assumed that the entire principal remains outstanding until maturity of the debt, that interest on the original principal is paid each period when due, and that the equal periodic payments into the sinking fund accumulate at compound interest to an amount which allows the repayment of the entire principal when it comes due. If S is the amount of the debt, n the number of payments into the sinking fund, and i the compound interest rate earned by the sinking fund, show that $S = Rs_{\overline{n}|i}$ or

$$R = S \cdot \frac{1}{s_{\overline{n}|i}}$$

if payments are made as often as interest is converted. (*Hint:* Interpret the sinking fund as an annuity whose amount is S and use Problem 3.)

9. Prove the following identity:

$$\frac{1}{s_{\overline{n}|i}} = \frac{1}{a_{\overline{n}|i}} - i.$$

(Note that this allows the calculation of the sinking fund factor, $1/s_{\overline{n}|i}$, by an easy subtraction from the corresponding entry in the tables for $1/a_{\overline{n}|i}$, rather than by a more difficult long division after finding $s_{\overline{n}|i}$.)

10. A debt of \$100 bearing interest at 5% compounded annually is to be repaid at the end of 10 years by accumulating a sinking fund for this time period. What is the annual sinking fund payment required if the fund can be invested to yield (*a*) 5%, (*b*) 6%? (Note: $1/a_{\overline{10}|.05} = 0.12950$, $1/a_{\overline{10}|.06} = 0.13587$.)

11. Find the annual payment to a depreciation fund, by the sinking fund method, for replacing a machine costing \$120 new, having a probable life of 10 years and a scrap value of \$20, if funds can be invested at 5% per annum. (See Problem 10.) If the *book value* at any time is the original cost minus the amount in the depreciation fund at that time, calculate the book value at the end of the first, fifth, and tenth payments. ($s_{\overline{5}|.05} = 5.5256$.)

12. Prove: If the sinking fund can be invested at the same rate of interest as that on the debt, the total periodic payment for interest and sinking fund is equal to the periodic payment required to amortize the debt.

13. An annuity in which the successive periodic payments increase (decrease) by a constant amount is called an increasing (decreasing) annuity. Let A_k denote the amount accumulated at the end of k conversion periods in an increasing annuity whose payments are $R, 2R, 3R, \cdots$ at the end of conversion

periods $1, 2, 3, \cdots$, with interest credited at the compound rate i per period. Show that

$$A_{k+1} = (1 + i)A_k + (k + 1)R \qquad k = 0, 1, 2, \cdots$$

with $A_0 = 0$. This is *not* a difference equation of the type in Theorem 2.7 since $(k + 1)R$ is not a constant independent of k. But show that

$$A_k = \frac{R}{i}[(1 + i)s_{\overline{k}|i} - k]$$

is a solution of both the difference equation and initial condition.

14. Consider the difference equation[9]

$$p_t = \left(1 - \frac{e}{D}\right)p_{t-1} + \frac{e}{D}P \qquad t = 1, 2, 3, \cdots,$$

where p_t denotes the price of bonds in period t, and e, D, and P are constants. Solve this equation (assuming p_0 specified) and thus show that the sequence $\{p_t\}$ converges to P if $e/D > 0$ and $e/D < 2$. Show further that the nature of the approach toward this equilibrium price depends upon the size of e/D, the sequence $\{p_t\}$ being monotone if $e/D \leq 1$, oscillatory if $e/D > 1$.

2.8 ECONOMIC DYNAMICS

We shall describe one of the classical models[10] used to study the growth of national income in an expanding economy. National income is made up of two components: consumption and investment (or desired accumulation). In this analysis, we are interested in following the variations in these quantities as time goes on. We shall suppose that time is divided into periods of equal length, say years, and introduce the functions Y, C, and I for national income, consumption, and investment respectively. These functions are thought of as defined with domain the set of t-values 0, 1, 2, $\cdots$ with Y_t, C_t, and I_t denoting their respective values at time t. The make-up of national income is then given by

$$(2.74) \qquad Y_t = C_t + I_t \qquad t = 0, 1, 2, \cdots.$$

Consumption expenditures are related to the national income level by the consumption function which we here assume to be linear. That is, we assume a straight-line graph of consumption plotted against income. The general form of such a relation is given by [cf. Example (d) p.12]

$$(2.75) \qquad C_t = c + mY_t \qquad t = 0, 1, 2, \cdots$$

[9] R. W. Clower, "Productivity, Thrift, and the Rate of Interest," *Economic Journal*, 64 (1954), 107–115.

[10] Actually we adapt Boulding's graphical analysis of what he calls "the Harrod-Domar-Hicks dynamic." See K. E. Boulding, "In Defense of Statics," *Quarterly Journal of Economics*, 69 (1955), 485–502, especially Section V.

where c and m are constants with the following geometric interpretations: the straight-line graph intersects the vertical consumption axis at the consumption value equal to c (i.e., when $Y_t = 0$, $C_t = c$), and the slope of the straight line is equal to m (i.e., $\Delta C_t = m \, \Delta Y_t$). The constant m is called the *marginal propensity to consume* (of this year's consumption related to this year's income). Note that we are assuming c and m independent of t so the relation between consumption and income is unchanging as t increases. We make the following restrictions on parameters c and m:

$$(2.76) \qquad\qquad c \geq 0 \qquad 0 < m < 1.$$

The second of these inequalities merely expresses the fact that any increase in income is in part, but not entirely, converted into an increase in consumption.

Suppose, for the sake of orientation, that we are in a period of full employment, with income at a level so that it is not all consumed, but a certain portion remains for investment. When invested, this portion will cause an increase in the capacity (or total national income) of the system and if full employment is to be maintained, investment expenditures will also have to grow. And this increase in investment will again increase capacity which in turn will increase investment, etc. The rate of growth of investment required to maintain full employment is Harrod's "warranted rate." Our aim is to determine this warranted rate and to describe the growth of both investment and national income with time.

We must, of course, make some statement about the precise manner in which the investment level influences the national income. We assume that there exists a constant, called the *growth factor*, denoted by r, for which

$$(2.77) \qquad\qquad \Delta Y_t = Y_{t+1} - Y_t = rI_t \qquad t = 0, 1, 2, \cdots.$$

The increase in capacity caused by unit investment is thus equal to r and we assume

$$(2.78) \qquad\qquad r > 0.$$

Relations (2.74) through (2.78) enable us to carry out the analysis of this dynamic "economy." Our first object is to derive difference equations satisfied by the functions Y and I.

Starting with (2.77) and using (2.74) and (2.75) in turn, we write

$$
\begin{aligned}
Y_{t+1} - Y_t &= rI_t \\
&= r(Y_t - C_t) \\
&= rY_t - r(c + mY_t)
\end{aligned}
$$

or

$$(2.79) \qquad Y_{t+1} = [1 + r(1 - m)] Y_t - rc \qquad t = 0, 1, 2, \cdots.$$

This is the linear first-order difference equation with constant coefficients satisfied by the national income function.

To find the corresponding equation for investment expenditures, we begin with (2.74) and write

$$I_{t+1} - I_t = (Y_{t+1} - C_{t+1}) - (Y_t - C_t)$$
$$= (Y_{t+1} - Y_t) - (C_{t+1} - C_t).$$

Now use (2.77) and (2.75) to simplify the right-hand side and obtain

$$I_{t+1} - I_t = rI_t - m(Y_{t+1} - Y_t)$$
$$= rI_t - mrI_t.$$

Thus

$$(2.80) \qquad I_{t+1} = [1 + r(1 - m)]I_t \qquad t = 0, 1, 2, \cdots.$$

Note that $r(1 - m) > 0$ from (2.76) and (2.78), so I is an increasing function of t. Investment grows at the constant rate $r(1 - m)$, this positive proportion of the investment I_t being added to I_t to yield I_{t+1}, the investment one period later. The difference equation (2.80) has the solution (with I_0 prescribed)

$$(2.81) \qquad I_t = [1 + r(1 - m)]^t I_0 \qquad t = 0, 1, 2, \cdots$$

and the sequence $\{I_t\}$ diverges to $+\infty$. (More realistic models with built-in ceilings and floors for investment and income levels have also been considered.[11])

The national income difference equation (2.79) can also be solved by using Theorem 2.7 with $A = 1 + r(1 - m)$, $B = -rc$, so that

$$(2.82) \qquad Y^* = \frac{B}{1 - A} = \frac{c}{1 - m}.$$

The solution (with Y_0 given) is

$$(2.83) \qquad Y_t = [1 + r(1 - m)]^t (Y_0 - Y^*) + Y^* \qquad t = 0, 1, 2, \cdots$$

so if $Y_0 > Y^*$, the sequence $\{Y_t\}$ also diverges to $+\infty$.

Harrod considered the special case in which $c = 0$, i.e., the consumption-income straight line passes through the origin. In this case, as may best

[11] For example, in J. R. Hicks, "Mr. Harrod's Dynamic Theory," *Economica*, New Series, *16* (1949), 106–121. This article, as well as the basic Harrod paper, has been reprinted in *Readings in Business Cycles and National Income*, A. H. Hansen and R. V. Clemence (eds.), W. W. Norton, New York, 1953.

be seen from the fact that the difference equations (2.79) and (2.80) assume precisely the same form, income and investment must grow at the *same* warranted rate of growth, given by the constant $r(1 - m)$, in order to maintain full employment.

PROBLEMS 2.8

1. Show that if $Y_0 = Y^*$, then $C_t = Y^*$ and $I_t = 0$ for all t.

2. In the simple multiplier model, two basic assumptions are made: (i) Y_t (total income) $= I_t$ (investment) $+ C_t$ (consumption), and (ii) consumption in any period is a linear function of the income of the *preceding* period, or $C_t = c + mY_{t-1}$, m being the marginal propensity to consume (of this year's consumption with respect to last year's income). Show that the income function satisfies the difference equation

$$Y_{t+1} = I_{t+1} + c + mY_t \qquad t = 0, 1, 2, \cdots$$

and find the solution of this equation assuming that investment is constant from period to period, say $I_t = i$ for all t. Assuming $0 < m < 1$, show that $\{Y_t\}$ is convergent with limiting value $Y^* = (i + c)/(1 - m)$. If $D_t = Y^* - Y_t$, show that $D_{t+1} = mD_t$ so that[12] "this system follows a path of simple exponential decline of D_t, the difference between Y_t and the equilibrium value $\cdots$."

3. Following Baumol's description[13] of the Harrod model, let Y_t, S_t, and I_t denote the income received, total savings, and entrepreneur's desired investment, respectively, of a community during period t. Translate the following assumptions into mathematical form: (i) the *realized* investment (savings) in any period is a constant proportion, s, of the income of that period, (ii) the entrepreneur's *desired* investment during any period is equal to a constant multiple, g, of the increase of the income of that period over the income of the preceding period, and (iii) the investors' desires are to be realized, i.e., savings in any period are equal to desired investment. From the equations expressing these assumptions, show that if $g \neq s$ and $c = g/(g - s)$, then

$$Y_{t+1} = cY_t \qquad t = 0, 1, 2, \cdots.$$

Solve this difference equation (with Y_0 prescribed) and then, assuming $Y_0 > 0$, $s > 0$, and $g > s$, show that income will be positive, nonoscillatory, and steadily increasing with time.

4. Consider the model of the preceding problem in the case when s is greater than g. Show that the income sequence $\{Y_t\}$ is damped oscillatory, oscillates finitely, or oscillates infinitely according as s is greater than, equal to, or less than $2g$.

---

[12] P. 490 of Boulding, cited in footnote 10.
[13] W. J. Baumol, *Economic Dynamics*, Macmillan, New York, 1951, pp. 42, 150; also "Formalisation of Mr. Harrod's Model," *Economic Journal*, 59 (1949), 625–629.

5. In Problem 3 modify assumption (i) so that the savings during any period are proportional to the income of the *preceding* period. With the other assumptions unchanged, show that now

$$Y_{t+1} = k Y_t \qquad t = 0, 1, 2, \cdots$$

where $k = (g + s)/g$. Thus prove that if s and g are both positive, Y_t steadily increases as t increases. (Note that the relative magnitudes of s and g are no longer crucial to the behavior of the sequence $\{Y_t\}$ and compare with Problems 3 and 4.)

6. In Problem 3, modify assumption (i) so that the savings during any period are proportional to the income anticipated in the *next* period and then show, with other assumptions unchanged, that income satisfies the second-order difference equation

$$Y_{t+2} = \frac{g}{s} Y_{t+1} - \frac{g}{s} Y_t \qquad t = 0, 1, 2, \cdots.$$

Suppose we have prescribed the initial values $Y_0 = 2$ and $Y_1 = 3$. Calculate the first ten terms of the sequence $\{Y_t\}$ if (a) $s = \frac{1}{10}$, $g = \frac{1}{2}$, and (b) $s = \frac{1}{10}$, $g = \frac{1}{5}$. Verify, in these particular cases, the following general result to be established in the next chapter: if $g \geq 4s > 0$, then $\{Y_t\}$ diverges to $+\infty$, but if $0 < g < 4s$, income oscillates.

7. In Problem 3, modify assumption (ii) to include both additional investment demand proportional to income and investment which is entirely independent of income, i.e., let $I_t = g(Y_t - Y_{t-1}) + KY_t + L$, where K and L are constants. Retaining the other assumptions, show that income now satisfies the difference equation

$$Y_{t+1} = \frac{-g}{s - g - K} Y_t + \frac{L}{s - g - K} \qquad t = 0, 1, 2, \cdots.$$

(Note that this reduces to the difference equation of Problem 3 in the extreme case $K = L = 0$.) Solve this equation to find

$$Y_t = \left(\frac{-g}{s - g - K} \right)^t \left(Y_0 - \frac{L}{s - K} \right) + \frac{L}{s - K} \qquad t = 0, 1, 2, \cdots.$$

Verify[14] that if $g > 0$ and $L > 0$, then whereas in Problem 3 $(Y_{t+1} - Y_t)/Y_{t+1}$ is constant $(= s/g)$, now this quantity, "the ratio of the increase of income to the original level of income, must increase as the latter increases."

8. (Expectational Price Cycles.[15]) If r_t denotes the amount of money demand for a commodity and p_t the price of the commodity in period t, assume a market equation of the form $p_t = Kr_t$, where K is a constant. Suppose further that "people project the trend of prices, so that rising prices lead to the expectation of further rise, and so to an increase in demand, while falling prices lead to an expectation of further fall, and so to a decrease in demand." To express this in

[14] P. 45 of Baumol's *Economic Dynamics*, cited in footnote 13.

[15] K. E. Boulding, *Economic Analysis*, 3rd ed., Harper, New York, 1955, p. 426.

simple form, assume that r_t is equal to some "normal" level r_0 plus a factor proportional to the increase of the price p_t over the price p_{t-1} of the preceding period, i.e., $r_t = r_0 + \alpha(p_t - p_{t-1})$. Derive the difference equation

$$p_{t+1} = \frac{-K\alpha}{1 - K\alpha} p_t + \frac{Kr_0}{1 - K\alpha} \qquad t = 0, 1, 2, \cdots,$$

and find its solution with p_0 prescribed. Show that if $K\alpha > 1$, then the sequence $\{p_t\}$ diverges to $+\infty$ (if $p_0 > Kr_0$) or diverges to $-\infty$ (if $p_0 < Kr_0$), but if $0 < K\alpha < 1$, then $\{p_t\}$ oscillates infinitely, oscillates finitely, or undergoes damped oscillations as $K\alpha$ is greater than, equal to, or less than $\frac{1}{2}$. In the case of damped oscillations, show that the equilibrium (limiting) price is Kr_0.

9. Suppose the market price, p_t, is determined by that period's supply, q_t, according to the relation

$$p_t = 2 - q_t.$$

At the end of period t, suppliers make an estimate of the next period's price. Let this expected price in period $t + 1$ be p_{t+1}^* and suppose production in period $t + 1$ is related to the expected price by the equation

$$q_{t+1} = p_{t+1}^* + 1.$$

(a) Show if $p_{t+1}^* = \frac{1}{2}$, then $p_{t+1}^* = p_{t+1}$, but that, in general, the actual and expected prices are unequal.

(b) Suppose some agency makes a public forecast[16] that the price in period $t + 1$ will be P and let

$$p_{t+1}^* = \alpha P + (1 - \alpha)p_t \qquad 0 \le \alpha \le 1.$$

(The constant α measures the extent to which suppliers have confidence in the public prediction as a better guide to the next period's price than the current price: $\alpha = 1$ means $p_{t+1}^* = P$ or perfect confidence, $\alpha = 0$ means $p_{t+1}^* = p_t$ or no confidence.) Show that the value of P for which the actual price in period $t + 1$ is equal to this public forecast is given by

$$P = \frac{1}{1 + \alpha} - \frac{1 - \alpha}{1 + \alpha} p_t.$$

2.9 INVENTORY ANALYSIS

A comprehensive mathematical model using difference equations has been presented[17] by L. A. Metzler in an analysis of inventory cycles. We shall outline only the beginning of this theory here. It is an example of

[16] E. Grunberg and F. Modigliani, "The Predictability of Social Events," *Journal of Political Economy*, 62 (1954), 465–478. See also W. J. Baumol, "Interactions Between Successive Polling Results and Voting Intentions," *Public Opinion Quarterly*, 21 (1957), 318–323, and the references there cited.

[17] L. A. Metzler, "The Nature and Stability of Inventory Cycles," *Review of Economics and Statistics*, 23 (1941), 113–129.

period analysis, in which time is divided into intervals of constant length (which may be thought of as years) and all functions are dated with the period at which they are to be evaluated.

Entrepreneurs produce consumer goods for two purposes: (1) for sales, and (2) for maintaining certain optimum inventory levels. With units of measurement appropriately chosen, we let u_t = number of units of consumers' goods produced for sale in period t, and s_t = number of units of consumers' goods produced for inventories in period t. We assume there is a constant noninduced net investment, denoted by v_0, in each period. The total income produced in period t is equal to the total production of consumers' goods plus net investment. If y_t = total income produced in period t, then

$$(2.84) \qquad y_t = u_t + s_t + v_0 \qquad t = 0, 1, 2, \cdots.$$

Producers planning their output for sales for any period do so at the beginning of that period and base their production plans on the sales of the preceding period. We assume actual sales of any period are a fraction, β, of the total income of that period. We further assume that planned output for period t is taken to be equal to actual sales of period $t - 1$, i.e.,

$$(2.85) \qquad u_t = \beta y_{t-1} \qquad t = 1, 2, 3, \cdots.$$

The constant β is the marginal propensity to consume (of this year's consumption with respect to last year's income). Note that (2.85) holds for t-values starting with $t = 1$ (rather than $t = 0$), since we are relating variables for period t and the preceding period $t - 1$.

Production for inventory stocks during any period is also decided at the beginning of each period and we assume that an attempt is made to maintain inventories at some constant (normal) level. That is, "businessmen will ordinarily attempt to replenish inventories depleted by an unforeseen rise in demand, or to reduce inventory accumulations resulting from unpredicted depressions."

To be definite, Metzler assumes that production of goods for inventories in any period t is equal to the difference between actual and anticipated sales of the preceding period $t - 1$. If this difference is $+10$, then actual sales exceeded expected sales by 10 units, inventory stocks were used to make up this difference, and production for inventory is planned to make up this loss of 10 units of stock. If this difference is -10, then inventory gained 10 units and production is planned so that in the next period these additional units will be used for sales. (Throughout we are assuming adequate inventories are maintained so that all differences between production and consumer demand can be met by inventory fluctuations rather than by price changes.)

Now anticipated sales of period $t - 1$ are u_{t-1}, which, by (2.85), is equal to βy_{t-2}. And by an assumption already made, actual sales of period $t - 1$ are β times the income of period $t - 1$, i.e., actual sales of period $t - 1$ are βy_{t-1}. Hence

$$(2.86) \qquad s_t = \beta y_{t-1} - \beta y_{t-2} \qquad t = 2, 3, \cdots.$$

It is now possible to derive a difference equation for the net income y. Starting with (2.84) and using (2.85) and (2.86), we obtain

$$y_t = \beta y_{t-1} + \beta y_{t-1} - \beta y_{t-2} + v_0$$

or

$$(2.87) \qquad y_t - 2\beta y_{t-1} + \beta y_{t-2} = v_0 \qquad t = 2, 3, \cdots.$$

If we wish to have our difference equation defined over the set of t-values beginning with $t = 0$, we rewrite (2.87) in the equivalent form

$$(2.88) \qquad y_{t+2} - 2\beta y_{t+1} + \beta y_t = v_0 \qquad t = 0, 1, 2, \cdots.$$

We recognize (2.88) as a second-order linear difference equation with constant coefficients. Having available only the theory of first-order difference equations, we postpone consideration of this particular model until Chapter 3. We shall then be able to show that income undergoes damped oscillatory movement toward some new equilibrium if an existing equilibrium is disturbed by an increase of noninduced investment. Metzler terms this "a pure inventory cycle" in the sense that it is produced by attempts to maintain inventories at some fixed level, i.e., by investment (or disinvestment) for inventory purposes.

Suppose, in the interest of obtaining a first-order difference equation, we abandon one of the assumptions already made and thus simplify our model. In fact, let us make the unrealistic assumption that all consumers' goods are produced for sale and no attempt is made by producers to maintain inventory levels. This is equivalent to assuming $s_t = 0$ rather than as given in (2.86). Equation (2.84) still stands and now reads

$$(2.89) \qquad y_t = u_t + v_0.$$

Production for sales is still governed by sales of the preceding period (which in turn are related to income of that period) by (2.85). Inserting this value of u_t in (2.89) yields the sought-for linear first-order difference equation with constant coefficients

$$(2.90) \qquad y_t = \beta y_{t-1} + v_0 \qquad t = 1, 2, \cdots.$$

We rewrite (2.90) as a difference equation over a set of t-values starting with $t = 0$. Then

$$(2.91) \qquad y_{t+1} = \beta y_t + v_0 \qquad t = 0, 1, 2, \cdots.$$

Theorem 2.7 may be applied with $A = \beta$, $B = v_0$, and

$$(2.92) \qquad y^* = \frac{v_0}{1 - \beta}$$

to yield the solution (with y_0 prescribed)

$$(2.93) \qquad y_t = \beta^t(y_0 - y^*) + y^* \qquad t = 0, 1, 2, \cdots.$$

With $0 < \beta < 1$, the sequence $\{y_t\}$ converges to the limiting income value y^*. The value y^* is that income sustainable by the noninduced investment v_0. If $y_0 > y^*$, the income decreases monotonically to y^*; if $y_0 < y^*$, then y increases steadily toward y^*; and if $y_0 = y^*$, then $y_t = y^*$ for all t.

To illustrate this result, we pose the following problem. With $\beta = 0.5$ and $v_0 = 100$, suppose we are in the equilibrium position in which the total income produced is 200 units. New investments are made so that v_0 is increased to 120 units. How does income respond to this injection of new investment? Let us first understand what is meant when we say that initially we are in an equilibrium position. We mean simply that as long as β and v_0 are unchanged, the level of income produced (in this case, 200) is maintained in all future time intervals. Note that, with $\beta = 0.5$ and $v_0 = 100$, the limiting value y^* as given by (2.92) is just 200. That this is the equilibrium value for income follows immediately from (2.93) since, as we have already observed, if $y_0 = y^*$, then $y_t = y^*$ for all later times.

This equilibrium is disturbed when v_0 is increased 20 units to its new value of 120. Let us write $t = 0$ as the first period for which $v_0 = 120$ and suppose that in this period the net income is the old equilibrium value $y_0 = 200$. Assume further that the inventory level at the end of this initial period is at 100 units. With these data given, we are now able to follow the variations in income, sales, and inventories as time moves forward in the sequence $0, 1, 2, \cdots$.

Production for sales in period 1 is denoted by u_1 and equals βy_0, or 100, in the specific case under consideration. Then $y_1 = u_1 + v_0 = 220$ and actual sales in period 1 are βy_1, or 110 units. Thus 10 more units are sold than produced and inventory stocks decrease by this amount to the level 90 by the end of period 1. Production in period 2 is $u_2 = \beta y_1 = 110$ units, $y_2 = u_2 + v_0 = 230$, and actual sales are $\beta y_2 = 115$ units, so inventories again decrease, but this time only 5 units to the new level 85. In this way, we can continue to compute the values of these functions for periods $3, 4, 5, \cdots$. The results up to $t = 5$ are summarized in Table 2.4.[18] We let i_t denote the level of inventories at the close of period t.

---

[18] Adapted from Table 1 of Metzler, cited in footnote 17.

We can find the value y_t for any t-value by using the solution (2.93) which in this specific example takes the form

$$y_t = (\tfrac{1}{2})^t(200 - 240) + 240$$

or

$$(2.94) \qquad y_t = -40(\tfrac{1}{2})^t + 240 \qquad t = 0, 1, 2, \cdots.$$

Since β is between 0 and 1 and y_0 is less than y^*, we know the sequence $\{y_t\}$ of income values increases steadily toward the limiting value $y^* = 240$.

TABLE 2.4

BEHAVIOR OF A SYSTEM WITH PASSIVE INVENTORY ADJUSTMENTS

$$\beta = 0.5, \ v_0 = 120, \ y_0 = 200, \ i_0 = 100$$

t	Production $u_t = \beta y_{t-1}$	Income $y_t = u_t + v_0$	Sales βy_t	Inventories i_t
0	–	200	100	100
1	100	220	110	90
2	110	230	115	85
3	115	235	117.5	82.5
4	117.5	237.5	118.75	81.25
5	118.75	238.75	119.375	80.625
.	.	.	.	.
.	.	.	.	.
.	.	.	.	.
∞	120	240	120	80

This is illustrated by the tabular values and the limit is explicitly indicated by the entry in the final row of the income column. Since the sequence $\{y_t\}$ converges with limit 240, we know (cf. 5 of Problems 2.6)

$$y_t \to 240 \quad \text{and} \quad y_{t-1} \to 240 \quad \text{as } t \to \infty.$$

Hence $u_t = \beta y_{t-1} \to 120$ and sales $= \beta y_t \to 120$ as $t \to \infty$. Thus sales increase from 100 to the limiting value 120 and there is a corresponding reduction of 20 in inventories. Inventories have the equilibrium level 80 as indicated in the final row of Table 2.4. The sequence $\{i_t\}$ of inventory values can also be studied directly by deriving a difference equation satisfied by the inventory function (cf. 3 and 4 of Problems 2.9). We return to the more general inventory cycle models in Chapter 3.

PROBLEMS 2.9

1. Why is it necessary to write the relation (2.86) with t-values starting with $t = 2$ rather than with $t = 0$ or $t = 1$?

2. Discuss the behavior of a system with passive inventory adjustments if the equilibrium positions (a) $\beta = 0.2$, $y_0 = 125$, $v_0 = 100$, and (b) $\beta = 0.8$, $y_0 = 500$, $v_0 = 100$, are disturbed by an increase in net investment, v_0, from 100 to 120 units. In each of the two cases, solve the difference equation (2.91), and draw up a table like Table 2.4. Is there a difference in the speed with which the limiting values are attained in the three cases $\beta = 0.2$, $\beta = 0.5$, and $\beta = 0.8$?

3. If i_t denotes the level of inventories at the close of period t, show that

$$i_t = i_{t-1} + s_t + u_t - \beta y_t.$$

With passive inventory adjustments, i.e., $s_t = 0$, use (2.89) and show that

$$i_t - i_{t-1} = y_t(1 - \beta) - v_0.$$

Now use (2.91) to eliminate y and obtain the equation

$$i_{t+1} - \beta i_t = i_t - \beta i_{t-1}.$$

Thus, show that $i_{t+1} - \beta i_t = c$, a constant, from which (with $\beta \neq 1$)

$$i_t = \beta^t \left(i_0 - \frac{c}{1 - \beta} \right) + \frac{c}{1 - \beta}.$$

4. With $i_0 = 100$, $i_1 = 90$, and $\beta = 0.5$, use the result of the preceding problem to show that $c = 40$ and

$$i_t = 20(0.5)^t + 80 \qquad t = 0, 1, 2, \cdots.$$

Calculate $i_2, i_3, i_4, \cdots$ from this formula and compare with the values in Table 2.4.

5. Check the tabular inventory entries made for the two cases of Problem 2 by calculating i_t from the solution found in Problem 3.

2.10 A PROBABILITY MODEL FOR LEARNING[19]

Suppose we wish to study how behavior changes under certain experimental conditions. We think, in particular, of a sequence of events starting with the perception of a stimulus, followed by the performance of a response (pressing bar, running maze, etc.), and ending with the occurrence of an environmental event (presentation of food, electric shock, etc.).

Behavior is measured by the probability, p, that the response will occur during some specified time interval after the sequence is initiated. The general idea is that p denotes the subject's level of performance and is increased or decreased after each occurrence of the response according as the environmental factors are reinforcing or inhibiting.

[19] As described by R. R. Bush and F. Mosteller, "A Mathematical Model for Simple Learning," *Psychological Review*, 58 (1951), 313–323. See also these authors' *Stochastic Models for Learning*, John Wiley, New York, 1955.

If we imagine an experiment in which a subject is repeatedly exposed to this sequence of events (stimulus-response-environmental event), we may divide the experiment into stages, each stage being a trial during which the subject is run through the sequence. The subject's level of performance is then a function of the trial number, denoted by n, and we let p_n be the probability of the response (during the specified time interval following the stimulus) in the nth trial run. The number p_0 will be taken as the initial value describing the disposition of the subject toward the response when he is first introduced to the experiment proper. The function p is then defined with domain the set of n-values $0, 1, 2, \cdots$. By calling p a probability we impose the norming

$$(2.95) \qquad 0 \leq p_n \leq 1 \qquad n = 0, 1, 2, \cdots,$$

which merely identifies the extremes of no response and certain response with the values 0 and 1 respectively.

We first assume that p_{n+1} depends on p_n only and not on the earlier values of the function p. In other words, the subject's performance in trial $n + 1$, although dependent on his level of behavior in the preceding trial (as measured by p_n), is independent of the past record of performance leading to trial n. This is referred to as the Markov property[20] of the model. Following Bush and Mosteller, we make the simplifying assumption that this dependence of p_{n+1} on p_n is a linear one, i.e., a straight line results when p_{n+1} is graphed as a function of p_n. The slope-intercept form of the equation of this straight line is [cf. Example (d), p. 12]

$$(2.96) \qquad p_{n+1} = a + mp_n \qquad n = 0, 1, 2, \cdots,$$

where a is the intercept (i.e., the value of p_{n+1} when p_n is 0) and m is the slope of the line (i.e., the amount by which p_{n+1} changes per unit increase of p_n).

For our purposes, it is more convenient to write this linear relation in the "gain-loss" form. We introduce the parameter b by the defining equation

$$(2.97) \qquad m = 1 - a - b$$

so relation (2.96) may be written as

$$(2.98) \qquad p_{n+1} = p_n + a(1 - p_n) - bp_n \qquad n = 0, 1, 2, \cdots.$$

[Note: if we know the values of a and m, then b is uniquely determined from (2.97); if a and b are known, then we may likewise determine m.

[20] See W. Feller, *An Introduction to Probability Theory and Its Applications*, Vol. 1, 2nd ed., John Wiley, 1957, p. 369.

Thus (2.96) and (2.98) are alternate but equivalent equations for the straight-line relation between p_n and p_{n+1}.]

If the subject's level of performance at trial number n is given by p_n, then $1 - p_n$ is the maximum possible increase in level and $-p_n$ is the maximum possible decrease in moving to trial $n + 1$. This follows since 1 and 0 are the largest and smallest values of p_{n+1}. Equation (2.98) may be translated by saying that the change in performance level, $\Delta p_n = p_{n+1} - p_n$, is proportional to the maximum possible gain and maximum possible loss. [It is for this reason that (2.98) was named the "gain-loss" form.] The constants of proportionality are a and b and we may therefore measure by the parameter a those environmental events which are reinforcing (e.g., presenting a reward) and by b those events which are inhibiting (e.g., punishing the subject).

Restrictions on a and b are imposed only in order to ensure that no matter what value p_n has, consistent with (2.95), the following value, p_{n+1}, will also be between 0 and 1 inclusive. If $p_n = 0$, then $p_{n+1} = a$, so we require

$$(2.99) \qquad\qquad 0 \le a \le 1.$$

If $p_n = 1$, then $p_{n+1} = 1 - b$, so $(1 - b)$ must be between 0 and 1. But this means we require

$$(2.100) \qquad\qquad 0 \le b \le 1.$$

We have proved that conditions (2.99) and (2.100) are *necessary* for p_{n+1} to be between the values 0 and 1. It is not hard to show that they are also *sufficient* conditions (cf. 1 of Problems 2.10). These restrictions are the only ones imposed on the parameters a and b appearing in the fundamental difference equation (2.98). Thus, $a = 0$ describes a situation in which no reward is given after the response occurs, $b = 0$ describes a no-punishment trial, and $a = b$ implies that the measures of reward and punishment are equal.

Quoting Bush and Mosteller, "we may now describe the progressive change in the probability of a response in an experiment such as the Graham-Gagné runway or Skinner box in which the same environmental events follow each occurrence of the response." Let us consider a specific example first. If $a = 0.4$ and $b = 0.1$, (2.98) becomes

$$p_{n+1} = p_n + 0.4(1 - p_n) - 0.1p_n$$

or

$$(2.101) \qquad p_{n+1} = 0.5p_n + 0.4 \qquad n = 0, 1, 2, \cdots.$$

If p_0 is assumed equal to 0.2, we may successively compute $p_1 = 0.5(0.2) + 0.4 = 0.5$, $p_2 = 0.5(0.5) + 0.4 = 0.65$, $\cdots$. To solve (2.101) and thus have p_n for every n, we use Theorem 2.7 with $A = 0.5$, $B = 0.4$, and

$$p^* = \frac{B}{1 - A} = \frac{0.4}{1 - 0.5} = 0.8$$

to obtain (with $p_0 = 0.2$)

$$p_n = (0.5)^n(0.2 - 0.8) + 0.8$$

or

(2.102) $p_n = -0.6(0.5)^n + 0.8 \qquad n = 0, 1, 2, \cdots.$

This solution shows the precise manner in which performance level varies with trial number. Since $0 < A < 1$ and $p_0 < p^*$, we know (see Table 2.2) that the sequence $\{p_n\}$ is monotone increasing with limit $p^* = 0.8$. Thus for repeated trials in which reward and punishment are weighted in the ratio $4 : 1$ (as with $a = 0.4$ and $b = 0.1$ in this particular example), the response and nonresponse probabilities, p_n and $1 - p_n$, approach limiting values p^* and $1 - p^*$ having this same ratio.

To return to the general case, we note that (2.98) may be rewritten in the standard form

(2.103) $p_{n+1} = (1 - a - b)p_n + a \qquad n = 0, 1, 2, \cdots.$

This we recognize as a linear first-order difference equation with constant coefficients. In fact, with the notation of Theorem 2.7, $A = 1 - a - b$, $B = a$, and

(2.104) $$p^* = \frac{B}{1 - A} = \frac{a}{a + b}$$

if a and b are not both 0 (in which case $A = 1$). We therefore have the solution

(2.105)

$$p_n = \begin{cases} (1 - a - b)^n \left(p_0 - \dfrac{a}{a+b} \right) + \dfrac{a}{a+b} & \text{if } a + b \neq 0 \\ p_0 & \text{if } a + b = 0. \end{cases} \qquad n = 0, 1, 2, \cdots$$

In view of (2.99) and (2.100), the constant $A = 1 - a - b$ is between -1 and 1, with the end points of this interval attained only if a and b are both 1 or both 0. If $a = b = 1$, then $A = -1$ and $\{p_n\}$ oscillates finitely between the two values p_0 and $1 - p_0$. But in all other cases the sequence $\{p_n\}$ converges, to the limit p_0 if $a = b = 0$, and to the limit p^* otherwise. If $0 < a + b < 1$, then $0 < A < 1$ and $\{p_n\}$ is monotone decreasing to p^* if $p_0 > p^*$, monotone increasing to p^* if $p_0 < p^*$, and always equal to p^* if $p_0 = p^*$. If $1 < a + b < 2$, then $-1 < A < 0$

and $\{p_n\}$ is a damped oscillatory sequence with limit p^*. The special case $a + b = 0$ yields a constant sequence (with value p_0); $a + b = 1$ produces a sequence each of whose elements is p^*.

Let us conclude with two special cases: (1) $a = 0$ and (2) $a = b$. Case (1) assumes that no reward is given after the response occurs. The difference equation (2.98) becomes

$$p_{n+1} = (1 - b)p_n$$

with the solution

$$p_n = (1 - b)^n p_0 \qquad n = 0, 1, 2, \cdots.$$

This is an equation which describes the steady decrease in response probability (as $n \to \infty$) from the initial probability p_0. By plotting p as a function of n we obtain a curve of *experimental extinction*. (Cf. the discussion in Chapter 0, Example 1.)

In case (2), $a = b$, the measures of reward and punishment are equal. If the extreme cases $a = b = 0$ and $a = b = 1$ are discounted, then the quantity $(1 - a - b)^n \to 0$ as $n \to \infty$ and (2.105) shows that $p_n \to p^*$, which is equal to 0.5 in case (2), $a = b$. That is, ultimately the response tends to occur (in the specified time interval after the stimulus is presented) in half the trials. The balancing of reward and punishment forces produces, in the long run, a corresponding symmetry in performance.

PROBLEMS 2.10

1. If $p_{n+1} = p_n + a(1 - p_n) - bp_n$ and $0 \leq p_n \leq 1$, show if $0 \leq a \leq 1$ and $0 \leq b \leq 1$, then $0 \leq p_{n+1} \leq 1$. [*Hint:* Establish the inequalities $p_n + a(1 - p_n) - bp_n \leq p_n + 1 \cdot (1 - p_n) - 0 = 1$ and $p_n + a(1 - p_n) - bp_n \geq p_n + 0(1 - p_n) - p_n = 0$.]

2. Using (2.98) with $a = 0.5$ and $b = 0.2$, calculate the values of p_{n+1} when $p_n = 0, 0.1, 0.2, \cdots, 0.9, 1$, and plot the straight-line graph measuring p_n along a horizontal and p_{n+1} along a vertical axis.

3. (*a*) Solve the difference equation (2.98) in the case $p_0 = 0.5$, $a = 0.1$, and $b = 0.4$, and show that when reward and punishment measures are in the ratio 1:4 the limiting probabilities of response and nonresponse are in the same ratio.

(*b*) Generalize this result by showing that when reward and punishment measures are in the ratio $a:b$ the limiting probabilities of response and nonresponse are in the same ratio.

4. With n measured along a horizontal and p_n along a vertical axis, plot graphs of the function p if (*a*) $a = 0$, $b = 0.2$, (*b*) $a = 0$, $b = 0.5$, and (*c*) $a = 0$, $b = 0.8$. Assume $p_0 = 0.5$ and show that the larger b is, the faster $p_n \to 0$.

5. Repeat the preceding problem with $p_0 = 0.5$ but (*a*) $a = 0.2$, $b = 0$, (*b*) $a = 0.5$, $b = 0$, and (*c*) $a = 0.8$, $b = 0$.

6. Suppose that $a = 0.3$ and $b = 0.2$. If $p_0 = 0.1$, calculate $p_1, p_2, p_3, \cdots$, show that $p_n \to 0.6$ as $n \to \infty$, and sketch the resulting *curve of acquisition*.

7. Consider the difference equation (2.98) with solution (2.105) in the special case $a = 1$, $b = 0$. Show that no matter what p_0 is prescribed, we have $p_1 = 1$ as well as $p_n = 1$ for all $n > 1$. (This is interpreted as insightful or one-trial learning.) How would you interpret the other extreme case in which $a = 0$, $b = 1$?

8. Let the operators Q (reward present) and E (reward absent) be defined by the equations

$$Qp = p + a(1 - p) - bp \qquad Ep = p - bp.$$

Start with p_0 and consider the sequence of probabilities $p_1 = Qp_0$, $p_2 = Ep_1$, $p_3 = Qp_2$, $p_4 = Ep^3, \cdots$ obtained by alternately applying the operators Q and E. Show that

$$p_2 = p_0 + a'(1 - p_0) - b'p_0$$

where

$$a' = a(1 - b) \qquad b' = 1 - (1 - b)^2.$$

Show further that if $P_n = p_{2n}$, then

$$P_{n+1} = P_n + a'(1 - P_n) - b'P_n \qquad n = 0, 1, 2, \cdots$$

and thus show that $\lim P_n = \lim p_{2n} = a'/(a' + b')$ as $n \to \infty$.

9. Generalize the preceding problem to the *fixed-ratio reinforcement*[21] schedule where only every kth response is rewarded. That is, starting with p_0, we obtain the sequence $p_1 = Qp_0$, $p_2 = Ep_1$, $p_3 = Ep_2, \cdots$, $p_k = Ep_{k-1}$, $p_{k+1} = Qp_k$, $p_{k+2} = Ep_{k+1}, \cdots, p_{2k} = Ep_{2k-1}, p_{2k+1} = Qp_{2k}, \cdots$. Show that

$$p_k = p_0 + a'(1 - p_0) - b'p_0$$

where

$$a' = a(1 - b)^{k-1} \qquad b' = 1 - (1 - b)^k.$$

Show further that if $P_n = p_{nk}$, then

$$P_{n+1} = P_n + a'(1 - P_n) - b'P_n \qquad n = 0, 1, 2, \cdots$$

and thus show $\lim p_{nk} = a'/(a' + b')$ as $n \to \infty$. If a and b are fixed, what happens to this limiting value as $k \to \infty$?

10. Denote[22] by $c(k, n)$ the probability that a relevant cue k has been conditioned at the beginning of the nth trial. Show that the hypothesis "on each trial of a given problem a constant proportion, θ, of unconditioned relevant cues become conditioned" may be translated by the equation

$$c(k, n + 1) = c(k, n) + \theta[1 - c(k, n)] \qquad n = 1, 2, 3, \cdots.$$

---

[21] Pp. 55, 66–68 of Bush and Mosteller's *Stochastic Models for Learning*, cited in footnote 19. Note that the operator denoted by E here is *not* the displacement operator of Chapter 1.

[22] F. Restle, "A Theory of Discrimination Learning," *Psychological Review*, 62 (1955), 11–19.

Note that this is a difference equation for the function c considered as a function of n. Solve this equation with the prescribed initial condition $c(k, 1) = 0$ and show that

$$c(k, n) = 1 - (1 - \theta)^{n-1} \qquad n = 1, 2, 3, \cdots.$$

11. Letting S_c denote the number of elements in a population of stimuli, R a class of behaviors, s_c the mean number of elements from S_c effective on any one trial, and x_k the expected number of elements from S_c which are conditioned to R in trial number k, Estes[23] derives the equation

$$x_{k+1} - x_k = s_c \frac{S_c - x_k}{S_c} \qquad k = 0, 1, 2, \cdots.$$

Solve this difference equation (with x_0 prescribed) to obtain

$$x_k = S_c - (S_c - x_0)\left(1 - \frac{s_c}{S_c}\right)^k.$$

12. In deriving expressions for spontaneous recovery and regression, Estes[24] introduces the following stimulus fluctuation model. A total set of stimuli, S^*, is subdivided into two sets, an available set S and an unavailable set S'. In each period of time, an element in S may either become a member of S' (with probability j) or remain in S (with probability $1 - j$). Similarly, an element originally in S' moves into S or stays in S' with probabilities j' and $1 - j'$ respectively. Let $f(t)$ denote the probability that a given element of S^* is in S at the end of period t. This element is in S at the end of period $t + 1$ if and only if one of the following occur: (a) it was in S at the end of period t and stayed there; (b) it was in S' at the end of period t and then moved into S. Thus show that

$$f(t + 1) = f(t) \cdot (1 - j) + [1 - f(t)] \cdot j' \qquad t = 0, 1, 2, \cdots.$$

Solve this difference equation to find (formula [1] in Estes' paper, cited in footnote 24)

$$f(t) = J - [J - f(0)](1 - j - j')^t$$

where $J = j'/(j + j')$.

13. Reinterpret the preceding problem, using the following new meanings for the symbols introduced there. Let set S^* be a panel[25] of individuals who are asked a certain question (capable of being answered by "yes" or "no"). Set S^* is divided into two sets: S (those individuals answering "yes") and S' (those answering "no"). In two successive panel polls in which the same question is

[23] W. K. Estes, "Toward a Statistical Theory of Learning," *Psychological Review*, 57 (1950), 94–107. Compare the result obtained in this problem with Estes' formula 3, p. 98, derived using differential equations.

[24] W. K. Estes, "Statistical Theory of Spontaneous Recovery and Regression," *Psychological Review*, 62 (1955), 145–154.

[25] T. W. Anderson, "Probability Models for Analyzing Time Changes in Attitudes," Chap. 1 in P. F. Lazarsfeld (ed.), *Mathematical Thinking in the Social Sciences*, Free Press, Glencoe, 1954, especially p. 30.

asked, an individual may move from S to S' or stay in S, and an individual in S' may move to S or stay in S'. That is, each person, whether answering "yes" or "no" in any poll, may keep or change his opinion in the next poll. The probability $f(t)$ is the probability that an individual answers "yes" in poll number t. Show that if $j = 0.1$ and $j' = 0.2$,

$$f(t) = \tfrac{2}{3} - [\tfrac{2}{3} - f(0)](0.7)^t \qquad t = 0, 1, 2, \cdots,$$

so that $f(t) \to \tfrac{2}{3}$ as $t \to \infty$.

14. Show that the difference equation for the function F_L

$$F_L(n + 1) = (1 - \theta_L)F_L(n) + \theta_L \pi \qquad n = 0, 1, 2, \cdots,$$

where π and θ_L are constants independent of n, has the solution[26]

$$F_L(n) = \pi - [\pi - F_L(0)](1 - \theta_L)^n.$$

The parameter θ_L is actually a probability so $0 \le \theta_L \le 1$. Show that as $n \to \infty$, $F_L(n) \to \pi$ unless $\theta_L = 0$.

15. Show that the difference equation[27]

$$A_n = (A_{n-1} + i)(1 - p) \qquad n = 1, 2, 3, \cdots$$

with $A_0 = 0$ has the solution (i and p being constants)

$$A_n = A^*[1 - (1 - p)^n] \qquad n = 0, 1, 2, \cdots$$

where $A^* = [i(1 - p)]/p$.

2.11 GEOMETRIC GROWTH

A number of problems, from diverse fields, may be placed in the following general form. Two related functions, X and Y, are given, with X assuming the values X_0, X_1, X_2, $\cdots$, and Y being defined for each of these X-values. We write Y_k for the value of Y when $X = X_k$. If X changes in value from X_k to X_{k+1}, i.e., if X undergoes the increment ΔX_k, then $\Delta Y_k = Y_{k+1} - Y_k$ denotes the corresponding change induced in the function Y. The relative change in Y is the absolute change divided by the original value of Y, i.e., $\Delta Y_k / Y_k$. We are interested here in studying pairs of such functions for which

$$(2.106) \qquad \Delta X_k = c \frac{\Delta Y_k}{Y_k} \qquad k = 0, 1, 2, \cdots,$$

i.e., the absolute change in X is proportional to the corresponding relative change in Y, the constant of proportionality being denoted by c.

---

[26] See C. J. Burke, W. K. Estes, and S. Hellyer, "Rate of Verbal Conditioning in Relation to Stimulus Variability," *Journal of Experimental Psychology*, *48* (1954), 153–161, especially pp. 154–155.

[27] R. B. Ammons, "Acquisition of Motor Skill: 1. Quantitative Analysis and Theoretical Formulation," *Psychological Review*, *54* (1947), 263–281.

Let us consider two examples.   (Others may be found in Problems 2.11.)

Example 1

In connection with his researches on the so-called St. Petersburg paradox in probability theory, Daniel Bernoulli in 1738 enunciated his principle of "moral expectation,"[28] the essential feature of which was the observation that the value (or utility) of a certain sum of money is very different according to the amount of money already owned.   He suggested a law by which an increment of gain has a utility proportional to this gain but inversely proportional to the amount owned before this gain.   Roughly speaking, the value of \$1 decreases steadily as one's wealth increases, its utility for one person being twice the utility of \$1 for another person twice as wealthy as the first.

Let us denote by w the function representing the wealth (say in monetary units) of a person and assume the sequence of values w_0, w_1, w_2, $\cdots$ as possible values for w.   The utility function will be denoted by U so that U_k is the utility associated with wealth w_k.   The Bernoulli hypothesis then reads

$$(2.107) \qquad \Delta U_k = c\, \frac{\Delta w_k}{w_k} \qquad k = 0, 1, 2, \cdots,$$

where $\Delta U_k = U_{k+1} - U_k$ is the increment of utility resulting from an increment $\Delta w_k = w_{k+1} - w_k$ of wealth.   It is customary to norm the utility function so that the utility associated with the wealth necessary for existence is zero.   If we assume w_0 to be this subsistence level of wealth, then this norming makes $U_0 = 0$.   Note that this utility-wealth equation may be obtained from (2.106) by identifying the variables X and Y with U and w respectively.

Example 2

The physiologist Ernst Heinrich Weber (1795–1878) observed that if two weights differ by a just noticeable amount, when the weights are increased their difference also must be proportionately increased in order that this difference remain just noticeable.   A weight must be increased by about one-fortieth of its amount for a change to be perceived.   Stimuli other than weight were seen to have the same property.   Illumination, for

---

[28] See the discussion and references cited in J. Stigler, "The Development of Utility Theory, II," *The Journal of Political Economy*, *58* (1950), 373–396.   A more mathematical exposition may be found in C. Jordan, "On Daniel Bernoulli's 'Moral Expectation' and on a New Conception of Expectation," *American Mathematical Monthly*, *31* (1924), 183–190.   For historical and critical comments on utility, see L. J. Savage, *The Foundations of Statistics*, John Wiley, New York, 1954, pp. 91–104.

example, must be increased by about one-hundredth of its amount for the change to be just noticeable.

Let us introduce the notation $S_0, S_1, S_2, \cdots$ for a sequence of stimulus values, with S_0 denoting a threshold stimulus which is just short of eliciting a response. Weber's law stems from his observations on just noticeable changes in stimuli and thus applies under the restrictive condition that the sequence $S_0, S_1, S_2, \cdots$ proceed in just noticeable steps; i.e., starting with S_0, we first notice a difference when the stimulus value is increased to S_1; starting at S_1, we must move to S_2 before noticing the change; etc. Under this assumption, *Weber's law* may be written in the form

$$(2.108) \qquad \frac{\Delta S_k}{S_k} = \frac{S_{k+1} - S_k}{S_k} = \text{a constant (independent of } k).$$

(This constant would be $\frac{1}{40}$ if S represents weight, $\frac{1}{100}$ if S denotes an illumination function, etc.)

Gustav Fechner (1801–1887) took Weber's equation and used it to define the magnitude of the sensation (or response) resulting from *any* given stimulus. He took the constant defined by (2.108) for just noticeable difference in S as the unit of sensation and considered the sequence $S_0, S_1, S_2, \cdots$ with no conditions imposed other than that S_0 be the threshold stimulus. If R_k denotes the sensation due to stimulus S_k, then Fechner assumed the increment in sensation, ΔR_k, due to the increment ΔS_k in stimulus, is proportional to the relative change in the stimulus; i.e., with c denoting the constant of proportionality,

$$(2.109) \qquad \Delta R_k = c \frac{\Delta S_k}{S_k} \qquad k = 0, 1, 2, \cdots.$$

This equation is known as the *Weber-Fechner law*.[29] It is the special case of (2.106) in which X and Y are interpreted as response (sensation) and stimulus respectively.

Returning to (2.106), we make the additional hypothesis that the sequence of X-values under consideration is a sequence of equally spaced values. That is, we assume the sequence $\{X_k\}$ is of the form $X_0, X_0 + h,$

---

[29] See E. G. Boring, *A History of Experimental Psychology*, 2nd ed., Appleton-Century-Crofts, New York, 1950, pp. 284–291; J. P. Guilford, *Psychometric Methods*, 2nd ed., McGraw-Hill, New York, 1954, pp. 37–42. For an interesting interpretation of this law, see M. L. De Fleur, "A Mass Communication Model of Stimulus Response Relationships: An Experiment in Leaflet Message Diffusion," *Sociometry*, *19* (1956), 12–25.

$X_0 + 2h$, $\cdots$ with any two consecutive values differing by the same positive constant h.   In general, we assume

(2.110) $X_k = X_0 + kh \qquad k = 0, 1, 2, \cdots$

with $h > 0$.   It follows that $\Delta X_k = h$ for every k so that (2.106) becomes

$$c\,\frac{\Delta Y_k}{Y_k} = h \qquad k = 0, 1, 2, \cdots.$$

If we simplify notation by writing

(2.111) $$\alpha = \frac{h}{c}$$

this equation reduces to

(2.112) $\Delta Y_k = \alpha Y_k$

or

(2.113) $Y_{k+1} = (1 + \alpha)Y_k \qquad k = 0, 1, 2, \cdots,$

a linear first-order difference equation with constant coefficients for the function Y. This equation, as shown by (2.112), is satisfied by any function Y whose increment (ΔY_k) is proportional to its magnitude (Y_k), or, equivalently, whose relative increment $(\Delta Y_k / Y_k)$ is a constant.

The solution of the difference equation (2.113) is easily found to be

(2.114) $Y_k = (1 + \alpha)^k Y_0 \qquad k = 0, 1, 2, \cdots.$

If Y is always positive (as in most applications), (2.112) and (2.113) show that $\alpha > 0$ if Y increases with k (i.e., $\Delta Y_k > 0$) and $-1 < \alpha < 0$ if Y decreases (i.e., $\Delta Y_k < 0$).   Now if $\alpha > 0$, we observe from (2.114) that the sequence $\{Y_k\}$ increases without bound (diverges to $+\infty$).   On the other hand, if $-1 < \alpha < 0$, $\{Y_k\}$ is steadily decreasing with limiting value 0.   A function Y with values Y_k satisfying (2.114) is said to *grow geometrically* (if $\alpha > 0$) or *decay geometrically* (if $-1 < \alpha < 0$).   It is sometimes also referred to as growth (or decay) at a constant rate.   [See 9 of Problems 2.11 and the compound interest law (2.67).]

Let us apply these results to the two examples used to illustrate (2.106). We have already assumed in Example 1 that $U_0 = 0$ so to satisfy (2.110) we need only suppose that $U_k = kh$, i.e., the utility value associated with wealth w_k is k times a unit of utility, denoted by the constant h.   Then (2.107) becomes

$$\Delta w_k = \frac{h}{c}\,w_k \qquad k = 0, 1, 2, \cdots$$

with solution given by (2.114),

$$(2.115) \qquad w_k = \left(1 + \frac{h}{c}\right)^k w_0 \qquad k = 0, 1, 2, \cdots.$$

Similarly, if we assume $\Delta R_k = h$ in the Weber-Fechner law, which amounts to considering sensations (or responses) which differ by the constant h, then (2.109) becomes

$$\Delta S_k = \frac{h}{c} S_k$$

with solution

$$(2.116) \qquad S_k = \left(1 + \frac{h}{c}\right)^k S_0.$$

Solution (2.115) shows the manner in which wealth (w_k) increases with increasing utility (i.e., as k increases) and (2.116) similarly indicates the geometric growth of stimulus with response. In both cases, we have growth, rather than decay, since h and c are both positive constants.

The utility-wealth and stimulus-response relations derived here are finite difference analogues of more usual differential relations. These latter lead to exponential growth curves rather than the geometric growth discussed here. In the next section, we shall derive limiting forms of (2.115) and (2.116) and thereby illustrate the close connection between difference and differential equations.

PROBLEMS 2.11

1. For the utility-wealth relation given by the Bernoulli hypothesis (2.107) with solution (2.115), assume $h = 0.5$, $c = 0.5$, and $w_0 = 1$. Since $U_k = kh$, we may plot a graph of this utility-wealth relation. Do this, plotting U_k on the vertical and w_k on the horizontal axis. [For example, when $k = 0$ we obtain the point ($w_0 = 1$, $U_0 = 0$); when $k = 1$ we have ($w_1 = 2$, $U_1 = 0.5$), etc.]

2. Sketch graphs as in the preceding problem, leaving the values of w_0 and h unchanged, but with (a) $c = 0.1$ and (b) $c = 1.5$. Observe that as c increases, the rate of increase of w decreases, the graphs approaching the constant (no-growth) horizontal line $w_k = w_0$.

3. Show that a relabeling of the axes in the preceding problems allows the graphs sketched there to be interpreted as graphs of the response-stimulus relations given in the solution (2.116) of the Weber-Fechner law.

4. Suppose[30] the members of a group are to be ranked according to the extent to which each participates in some group task. Let p_k denote the value of some

---

[30] F. F. Stephan and E. G. Mishler, "The Distribution of Participation in Small Groups: An Exponential Approximation," *American Sociological Review*, *17* (1952), 598–608; reprinted in A. P. Hare, E. F. Borgatta, and R. F. Bales (eds.), *Small Groups* (*Studies in Social Interaction*), Knopf, New York, 1955.

quantitative (actual or estimated) measure of participation for an individual in rank k, and let

$$p_k = ar^k \qquad k = 0, 1, 2, \cdots,$$

where a and r are positive constants and $r < 1$. (a) Show that a is the measure of participation for an individual in rank 0. (b) Show that $p_{k+1} = rp_k$ so that r is the ratio of participation for any rank to the participation for the next lower rank. (c) Show that the function p decreases geometrically.

5. Denote by N_t the number of individuals in a certain population at the end of year t ($t = 0, 1, 2, \cdots$). Show that the difference equation $\Delta N_t = N_{t+1} - N_t = 0.02N_t$ expresses the assumption that the population undergoes a 2% relative increase in size per year, or, equivalently, that the increase in size in any year is proportional to the size at the beginning of that year, the constant of proportionality being 0.02. Solve this difference equation to obtain

$$N_t = (1 + 0.02)^t N_0 \qquad t = 0, 1, 2, \cdots,$$

where N_0 denotes the initial population size. Note that the population grows geometrically. To find the number of years required for doubling the population size, put $N_t = 2N_0$ and solve for t. Compare with 1(b) of Problems 2.7.

6. Suppose the natural increase in population size of the preceding problem is offset by a constant annual loss of ten individuals due to adverse conditions. Show that

$$N_{t+1} = (1.02)N_t - 10 \qquad t = 0, 1, 2, \cdots$$

and solve this difference equation to find

$$N_t = (1.02)^t(N_0 - 500) + 500 \qquad t = 0, 1, 2, \cdots.$$

To find the number of years required for doubling the population size put $N_t = 2N_0$ and solve for t. Show that this value of t satisfies the equation

$$(1.02)^t = \frac{2N_0 - 500}{N_0 - 500}$$

and thus depends upon the initial population size N_0. (Cf. the preceding problem.) Show that this value of t is 41 years if $N_0 = 2500$. (Note: $(1.02)^{41} = 2.25$ correct to two decimal places.)

7. Suppose the natural increase in population size of Problem 5 is augmented by an annual immigration of ten individuals. Formulate the appropriate difference equation for N and show that

$$N_t = (1.02)^t(N_0 + 500) - 500 \qquad t = 0, 1, 2, \cdots.$$

Show that approximately 31 years are required for an initial population of 2500 individuals to double itself. (Note: $(1.02)^{30} = 1.81$, $(1.02)^{31} = 1.85$.) Does this number of years depend upon the initial population size?

8. Suppose[31] a well produces q_n barrels of oil per day during time interval n ($n = 0, 1, 2, \cdots$) and let there be a constant percentage relative decline in

[31] E. Schweyer, *Process Engineering Economics*, McGraw-Hill, New York, 1955, p. 41.

production assumed to occur only at the end of each time interval. If the numerical value of this fixed percentage equals $100f\%$, show that $\Delta q_n = -fq_n$ so that q decreases geometrically according to the formula $q_n = (1 - f)^n q_0$.

9. According to Boulding,[32] the profit-making process may be thought of as the process by which the total net worth of the economy grows. Suppose the net worth at the end of year t is denoted by W_t $(t = 0, 1, 2, \cdots)$ and assume net worth has a constant rate of growth of 5% per year. Show that the function W satisfies the difference equation

$$W_{t+1} - W_t = 0.05 W_t \qquad t = 0, 1, 2, \cdots$$

and solve to find

$$W_t = (1.05)^t W_0 \qquad t = 0, 1, 2, \cdots.$$

Sketch a graph of the growth of net worth assuming $t = 0$ is a base year for which $W_0 = 1$. Measure t along the horizontal and W_t along the vertical axis.

10. Suppose[33] time is divided into practice periods of constant length and that the level of attainment in certain learning experiments is measured by Y_k, the number of successful acts performed in practice period k $(k = 0, 1, 2, \cdots)$. If L is a constant measuring the limit of attainment in terms of number of successful acts performed per period, then $y_k = L - Y_k$ is defined as the margin of attainment in period k. If the relative change of the margin of attainment is a constant decrease of magnitude $1/R$, show that $\Delta y_k = -(1/R)y_k$ and solve the difference equation to obtain

$$y_k = y_0 \left(1 - \frac{1}{R}\right)^k \qquad k = 0, 1, 2, \cdots.$$

With $y_0 = 10$ and $R = 5$, calculate $y_1, y_2, \cdots, y_7$, graph the margin of attainment as a function of the number of practice periods, and show that y decreases geometrically.

*2.12 APPROXIMATING A DIFFERENTIAL EQUATION BY A DIFFERENCE EQUATION

In view of the close relation between the finite difference operator Δ and the differential operator D (see Section 1.7) it should not be surprising to find many close connections between difference equations and differential equations. In this section we illustrate this point by showing that it is possible to find the solution of a differential equation as an appropriate limit of the solution of a corresponding difference equation. We restrict our attention here to the linear first-order differential equation with constant coefficients. Let y be a function, defined

---

[32] P. 844 of reference cited in footnote 15, page 97.

[33] H. J. Ettlinger, "A Curve of Growth Designed to Represent the Learning Process," *Journal of Experimental Psychology*, 9 (1926), 409–414, especially footnote on p. 414.

for real values x in some interval $a \leq x \leq b$, satisfying such a differential equation.   Then

$$(2.117) \qquad Dy(x) = \frac{dy(x)}{dx} = Ay(x) + B \qquad a \leq x \leq b,$$

where A and B are arbitrary constants with $A \neq 0$. We suppose the value of y at $x = a$ is prescribed, i.e., we seek a function y satisfying the differential equation (2.117) and the initial condition

$$(2.118) \qquad\qquad\qquad y(a) = y_0,$$

where y_0 is some given constant.

To approximate this differential equation by som' difference equation first requires that we replace the continuous interval $a \leq x \leq b$ by a discrete set of x-values over which the difference equation may be defined. Let us choose any positive integer n and divide the interval from a to b into n equal parts, each of length

$$(2.119) \qquad\qquad\qquad h = \frac{b - a}{n},$$

by the points of subdivision

$$(2.120) \quad x_0 = a, \quad x_1 = a + h, \quad x_2 = a + 2h, \cdots, \quad x_n = a + nh = b.$$

Let us further simplify the notation by writing

$$(2.121) \qquad\qquad y_k = y(x_k) = y(x_0 + kh).$$

Recalling that

$$(2.122) \qquad\qquad Dy(x_k) = \lim_{h \to 0} \frac{\Delta y_k}{h},$$

it seems reasonable to replace the differential equation (2.117) by the difference equation

$$(2.123) \quad \frac{\Delta y_k}{h} = Ay_k + B \qquad\qquad k = 0, 1, 2, \cdots, n - 1$$

or[34]

$$(2.124) \qquad y_{k+1} = (1 + Ah)y_k + Bh \qquad k = 0, 1, 2, \cdots, n - 1.$$

The initial condition (2.118) means that the value y_0 is prescribed.

---

[34] When $k = n - 1$ the difference equation involves the value y_n. Since the function y may not be defined for values of x beyond $x_n = b$, we cannot define the difference equation beyond $k = n - 1$.

The difference equation (2.124) is a first-order linear equation with constant coefficients and can easily be solved by the application of Theorem 2.7.   We have

$$y^* = \frac{Bh}{1 - (1 + Ah)} = -\frac{B}{A}$$

and therefore

$$(2.125) \quad y_k = (1 + Ah)^k \left(y_0 + \frac{B}{A}\right) - \frac{B}{A} \qquad k = 0, 1, 2, \cdots, n - 1.$$

Proceeding formally, without regard for rigorous proof, it appears that starting with the difference equation (2.123), if we let

$$(2.126) \qquad h \to 0 \quad \text{and} \quad x_k = x_0 + kh \to x,$$

then (2.123) becomes the differential equation (2.117).   It is therefore of some interest to inquire whether the difference equation solution approaches a solution of the differential equation under these same limiting conditions. We now show that this actually is the case.

$$y(x) = \lim y_k$$

$$= \lim \left[(1 + Ah)^k \left(y_0 + \frac{B}{A}\right) - \frac{B}{A}\right]$$

$$(2.127) \qquad = \left(y_0 + \frac{B}{A}\right) \lim (1 + Ah)^k - \frac{B}{A}.$$

But

$$\lim (1 + Ah)^k = \lim (1 + Ah)^{\frac{x - x_0}{h}}$$

which we write in the form

$$\lim [(1 + \varepsilon)^{\frac{1}{\varepsilon}}]^{A(x - x_0)}$$

with ε written in place of Ah.   Let us now recall that the number e, approximately equal to 2.71828, is defined as a limit:

$$e = \lim_{\varepsilon \to 0} (1 + \varepsilon)^{\frac{1}{\varepsilon}}.$$

Under the limiting conditions in (2.126), $h \to 0$ and since A is a constant, $Ah = \varepsilon$ also approaches 0.   Hence

$$\lim [(1 + \varepsilon)^{\frac{1}{\varepsilon}}]^{A(x - x_0)} = e^{A(x - x_0)}$$

and from (2.127),

$$(2.128) \qquad y(x) = \lim y_k = \left(y_0 + \frac{B}{A}\right) e^{A(x - x_0)} - \frac{B}{A}.$$

That the function y whose value at x is given by (2.128) is actually a solution of the differential equation (2.117) can be proved by taking its derivative and showing (2.117) is satisfied for every x in the interval $a \leq x \leq b$. We leave this calculation for the reader. Note also that (2.128) satisfies the initial condition (2.118). In fact, the function given by (2.128) is the unique solution of the differential equation which satisfies this condition.

The special case in which $B = 0$ is of particular interest because of important applications. Then

$$(2.129) \qquad Dy(x) = \frac{dy(x)}{dx} = Ay(x)$$

or the instantaneous rate of change of the function y with respect to x is proportional to the function y. The corresponding difference equation approximation is obtained from (2.124). It is

$$(2.130) \qquad y_{k+1} = (1 + Ah)y_k \qquad k = 0, 1, 2, \cdots, n - 1$$

with solution

$$(2.131) \qquad y_k = y_0(1 + Ah)^k.$$

Under the limiting conditions already described, the corresponding solution of the differential equation (2.129) is given by the exponential function

$$(2.132) \qquad y(x) = y_0 e^{A(x - x_0)}$$

with y_0 denoting the prescribed value of y when $x = x_0$. If y assumes positive values (as in most applications), y is said to undergo *exponential growth* (if $A > 0$) or *exponential decay* (if $A < 0$) as x increases from $x = x_0$. Exponential growth is thus seen to be the continuous analog of the discontinuous geometric growth discussed in the preceding section.

One often encounters formulations of "laws" in the framework of this continuous differential equation with an exponential solution rather than in the corresponding discrete difference equation form. For example, the utility-wealth relation (2.115)

$$w_k = \left(1 + \frac{h}{c}\right)^k w_0$$

is of the form (2.131) with A replaced by $1/c$. In the limiting exponential form (2.132) we identify y and x with the new symbols w and U to obtain (since $U_0 = 0$)

$$(2.133) \qquad w(U) = w_0 e^{\frac{U}{c}}.$$

Recalling the relation between the natural logarithmic function (denoted by ln) and the exponential function, we may write (2.133) in the equivalent and more usual form in which the utility of a given wealth is explicitly given as

$$U = c \ln \frac{w}{w_0}.$$

Similarly, we find the limiting form of the Weber-Fechner law as

$$S = S_0 e^{\frac{R}{c}},$$

or equivalently

$$R = c \ln \frac{S}{S_0}.$$

These examples merely illustrate some of the possibilities involved in using difference equations as approximations for differential equations. This has proved to be an important analytic tool in the numerical solution of differential equations whose explicit solutions are difficult to obtain. But the reader should take note of the fact that problems in this field are generally difficult. In addition to the question of choosing an appropriate difference equation approximation for a given differential equation, there is the troublesome problem of analyzing the propagation of errors in such numerical techniques.[35]

[35] For a brief indication of some possible difficulties in approximating a differential equation by a difference equation, see W. T. Martin and E. Reissner, *Elementary Differential Equations*, Addison-Wesley, Cambridge, Mass., 1956, pp. 205–208. For error analysis in the numerical solution of differential equations, consult one of the treatises on numerical analysis, e.g., F. B. Hildebrand, *Introduction to Numerical Analysis*, McGraw-Hill, New York, 1956, Chap. 6.

3

Linear Difference Equations with Constant Coefficients

3.1 SOME BASIC THEOREMS

A difference equation over the set of k-values $0, 1, 2, \cdots$ is said to be *linear* (see Definition 2.2) if it is of the form

$$(3.1) \quad f_0(k)y_{k+n} + f_1(k)_{k+n-1} + \cdots + f_{n-1}(k)y_{k+1} + f_n(k)y_k = g_k$$

where $f_0, \cdots, f_n$, and g are functions of k defined for $k = 0, 1, 2, \cdots$. If the coefficients $f_0, \cdots, f_n$ are each constant functions, the difference equation is linear *with constant coefficients*. (The function g is arbitrary and may, but need not, be constant.) If both f_0 and f_n are nonzero constants, (3.1) is a linear difference equation with constant coefficients and of *order n*. For example, the linear difference equations

$$\begin{align}
2y_{k+1} - y_k &= 6 \\
(3.2) \qquad 3y_{k+2} + 2y_{k+1} + y_k &= 3^k \\
y_{k+3} - y_k &= k
\end{align}$$

have constant coefficients and are of orders 1, 2, and 3 respectively.

When the leading coefficient (i.e., the coefficient of y_{k+n}) is not 1, it is convenient to divide the equation through by this leading coefficient.

The third of the equations in (3.2) is already in this form, but the first two become

(3.3)
$$y_{k+1} - \tfrac{1}{2}y_k = 3$$
$$y_{k+2} + \tfrac{2}{3}y_{k+1} + \tfrac{1}{3}y_k = 3^{k-1}.$$

In general, if (3.1) is of order n, then f_0 is different from zero and we may divide by f_0 and rename our coefficients $a_1 = f_1/f_0$, $a_2 = f_2/f_0$, $\cdots$, $a_n = f_n/f_0$, $r_k = g_k/f_0$. We obtain

(3.4) $$y_{k+n} + a_1 y_{k+n-1} + \cdots + a_{n-1}y_{k+1} + a_n y_k = r_k.$$

(Since f_n is also different from zero, we know $a_n \neq 0$.) It is in this form that we shall develop the theory of linear difference equations with constant coefficients. *We assume throughout this chapter that the coefficients $a_1, a_2, \cdots, a_n$ are constants (with $a_n \neq 0$) and that the right-hand term r is an arbitrary function defined for $k = 0, 1, 2, \cdots$. Unless the contrary is explicitly indicated, we assume that (3.4) is defined over this same set of k-values.*

With this notation, we have the following general forms for the linear difference equations with constant coefficients of orders 1, 2, and 3:

(3.5) $$y_{k+1} + a_1 y_k = r_k$$

(3.6) $$y_{k+2} + a_1 y_{k+1} + a_2 y_k = r_k$$

(3.7) $$y_{k+3} + a_1 y_{k+2} + a_2 y_{k+1} + a_3 y_k = r_k.$$

Thus, with $a_1 = -\tfrac{1}{2}$ and $r_k = 3$, (3.5) becomes the first of the two equations in (3.3), whereas the second requires that we put $a_1 = \tfrac{2}{3}$, $a_2 = \tfrac{1}{3}$, and $r_k = 3^{k-1}$ in (3.6). The final equation in (3.2) is the special case of (3.7) in which $a_1 = a_2 = 0$, $a_3 = -1$, and $r_k = k$.

Equation (3.4) with the right-hand side identically zero,

(3.8) $$y_{k+n} + a_1 y_{k+n-1} + \cdots + a_{n-1}y_{k+1} + a_n y_k = 0,$$

is said to be the *homogeneous* (or *reduced*) difference equation corresponding to the *nonhomogeneous* (or *complete*) difference equation (3.4).

The properties of linear difference equations (whether with constant coefficients or not) which account for the satisfactory state of their theory are embodied in the following results.

THEOREM 3.1. *If $y^{(1)}$ and $y^{(2)}$ are any two solutions of the linear homogeneous difference equation (3.8), then $C_1 y^{(1)} + C_2 y^{(2)}$ is also a solution for arbitrary constants C_1 and C_2.*

PROOF. We write out the details of the proof in the case $n = 2$, i.e., for the homogeneous linear difference equation of order 2,

$$(3.9) \qquad\qquad y_{k+2} + a_1 y_{k+1} + a_2 y_k = 0$$

corresponding to the complete equation (3.6). We must prove that if $y^{(1)}$ and $y^{(2)}$ each satisfy (3.9), then $C_1 y^{(1)} + C_2 y^{(2)}$ likewise does so. Thus we must show that

$$(C_1 y_{k+2}^{(1)} + C_2 y_{k+2}^{(2)}) + a_1(C_1 y_{k+1}^{(1)} + C_2 y_{k+1}^{(2)}) + a_2(C_1 y_k^{(1)} + C_2 y_k^{(2)}) = 0$$

is an identity, i.e., is true for $k = 0, 1, 2, \cdots$.

We rewrite the left-hand side of this equation by collecting values of $y^{(1)}$ and $y^{(2)}$ and obtain

$$C_1(y_{k+2}^{(1)} + a_1 y_{k+1}^{(1)} + a_2 y_k^{(1)}) + C_2(y_{k+2}^{(2)} + a_1 y_{k+1}^{(2)} + a_2 y_k^{(2)}).$$

That this sum is indeed 0 is clear, since each sum in parentheses is 0 by virtue of the assumption that both $y^{(1)}$ and $y^{(2)}$ are solutions of (3.9).

We note that this same direct proof, although more cumbersome to write, may be given for the general difference equation of order n in (3.8). It is also important to understand that the proof in no way uses the fact that the coefficients are constants, so that this is indeed a property of *linear* equations, whether with constant coefficients or not.

We add a corollary whose proof is left for the reader (see 6 of Problems 3.1). It asserts that we may add more than two solutions and still have a solution of the difference equation. By a *finite linear combination of solutions* we mean the sum of a finite number of solutions, each multiplied by an arbitrary constant, i.e.,

$$C_1 y^{(1)} + C_2 y^{(2)} + C_3 y^{(3)} + \cdots + C_n y^{(n)}.$$

Then the following extension of Theorem 3.1 can be proved.

COROLLARY. *Any finite linear combination of solutions of the linear homogeneous difference equation* (3.8) *is also a solution of this difference equation.*

We are interested in finding solutions of the complete equation. In fact, our aim is to solve this nonhomogeneous difference equation, i.e., find *all* its solutions. The theorem just proved together with the following one are key results in this direction.

THEOREM 3.2. *If Y is a solution of the homogeneous equation* (3.8) *and y^* is a solution of the complete equation* (3.4), *then $Y + y^*$ is a solution of* (3.4).

PROOF.   We write out the details of the proof for the difference equation (3.6) of order 2.   The method of proof is identical in the general case.   By hypothesis,

$$Y_{k+2} + a_1 Y_{k+1} + a_2 Y_k = 0$$

and

$$y_{k+2}^* + a_1 y_{k+1}^* + a_2 y_k^* = r_k.$$

Now add these equations to obtain

$$(Y_{k+2} + y_{k+2}^*) + a_1(Y_{k+1} + y_{k+1}^*) + a_2(Y_k + y_k^*) = r_k,$$

which is the desired conclusion.

This theorem points to a possible mode of attack in seeking to solve the complete difference equation (3.4).   It may (and indeed does) turn out to be sound strategy to break the problem up into two parts.   Instead of finding *all* the solutions of the complete equation, we settle for any *one*, which we name y^*.   Inasmuch as there are, in general, other solutions, we call y^* a *particular solution* of the complete equation.   This particular solution does not solve our problem.   We want the *general solution* of the complete equation, by which we mean a formula which includes all solutions as special cases.

Now turn to the homogeneous equation.   Theorem 3.2 tells us that if Y is a solution of this homogeneous equation, then $Y + y^*$ will be *a* solution of the complete equation.   Let us look at this from a different point of view.   With y^* already found, is there a choice of the function Y for which $Y + y^*$ will include *all* solutions of the complete equation? The best we can hope to do is let Y denote the general solution of the homogeneous equation, i.e., let Y be of sufficient generality to yield all solutions of this equation.   It turns out that our best is good enough. As a matter of fact, we shall be able to prove that the general solution of the complete equation is obtainable as the sum of the general solution of the homogeneous equation and any particular solution of the complete equation.

To illustrate the use of this methodology, let us solve the simplest linear equation, that of first order with constant coefficients,

$$y_{k+1} + a_1 y_k = r_k \qquad k = 0, 1, 2, \cdots.$$

We first consider the corresponding homogeneous equation

$$y_{k+1} + a_1 y_k = 0,$$

which is a special case of the difference equation solved in the preceding chapter.   In fact, from the corollary to Theorem 2.2 we know (with $A = -a_1$, $B = 0$) that the function Y whose value at k is

$$Y_k = C(-a_1)^k,$$

with C an arbitrary constant, is the general solution of this homogeneous difference equation.

Let us now suppose that by some means one particular solution, y^*, of the complete equation has been discovered. By Theorem 3.2, the sum $C(-a_1)^k + y_k^*$ is also a solution of the complete equation. But have we found *all* the solutions in this way? The answer is affirmative, as we now prove.

THEOREM 3.3. *Consider the linear first-order difference equation with constant coefficients*

$$y_{k+1} + a_1 y_k = r_k.$$

(*a*) *The function* Y *given by*

$$Y_k = C(-a_1)^k,$$

with C *an arbitrary constant, is the general solution of the corresponding homogeneous equation*

$$y_{k+1} + a_1 y_k = 0.$$

(*b*) *If* y^* *is any particular solution of the complete equation, then* $Y + y^*$ *is the general solution of the complete equation. That is, if* y *is any solution of the complete difference equation, there is a value of the constant* C *for which*

$$y_k = C(-a_1)^k + y_k^*.$$

PROOF. As we have already observed, (*a*) is a consequence of the corollary to Theorem 2.2. To prove (*b*), let y be any solution of the complete equation. By Theorem 3.2, the function $Y + y^*$ is a solution of the complete equation for every value of the constant C. We must prove that a value of C exists for which $Y + y^*$ and y are the same solution. But two solutions of a linear first-order difference equation which have the same value when $k = 0$ are identical. This was the substance of the uniqueness theorem of Section 2.3. Thus, our proof will be complete if C can be determined so that

$$y_0 = C(-a_1)^0 + y_0^*.$$

But $(-a_1)^0 = 1$ so

$$C = y_0 - y_0^*$$

is the required value.

Examples

(*a*) Consider the difference equation

(3.10) $$y_{k+1} - 2y_k = 5.$$

The homogeneous equation

$$(3.11) \qquad y_{k+1} - 2y_k = 0$$

has the general solution Y given by

$$(3.12) \qquad Y_k = C \cdot 2^k.$$

Since we need *any* particular solution of the complete equation, we try finding one among the simplest of functions, the constants. Suppose we put $y_k{}^* = A$ and try to determine the constant A so that y^* will satisfy the complete equation. We must have

$$A - 2A = 5$$

so that

$$y_k{}^* = A = -5.$$

Thus the general solution of (3.10) is $Y + y^*$ or

$$(3.13) \qquad y_k = C \cdot 2^k - 5.$$

If we are asked to find that solution of (3.10) for which

$$(3.14) \qquad y_0 = 4,$$

we determine the particular value of C in (3.13) for which this initial condition is satisfied. Put $k = 0$ in (3.13) to find

$$y_0 = C - 5.$$

In view of (3.14) we take $C = 9$. Thus, if

$$y_k = 9 \cdot 2^k - 5,$$

y is the solution satisfying both the difference equation (3.10) and the initial condition (3.14). This is, of course, the same result as that obtainable by using Theorem 2.7.

(*b*) Now consider the difference equation

$$(3.15) \qquad y_{k+1} - 2y_k = k + 1.$$

Since the corresponding homogeneous equation is again (3.11), we have the general solution (3.12) as in the preceding example. The only problem is to find y^*. It is now impossible to find a constant function which will satisfy (3.15). For if we again put $y_k{}^* = A$, we find

$$A - 2A = k + 1.$$

Thus $A = -k - 1$ is not constant, but depends upon k. So we optimistically try to find y^* among the slightly more complicated functions of the form

$$y_k{}^* = Ak + B,$$

where A and B are constants. (This method of finding particular solutions is known as the method of undetermined coefficients and is treated in Section 3.4.) Now

$$y_{k+1}^* = A(k + 1) + B$$

so if y^* is to satisfy (3.15), we require that

$$A(k + 1) + B - 2(Ak + B) = k + 1.$$

Collecting all terms on the left,

$$(-A - 1)k + (A - B - 1) = 0.$$

This will be an identity, i.e., true for every value of k, if the terms in parentheses are 0. Then $A = -1$ and $B = -2$, so that

$$y_k^* = -k - 2.$$

Thus the general solution of the complete equation (3.15) is given by

(3.16) $$y_k = C \cdot 2^k - k - 2.$$

(Note that this result could not be obtained from Theorem 2.7, since (3.15) does not have a *constant* right-hand term.)

The particular solution of (3.15) which satisfies the initial condition (3.14) is found by putting $k = 0$ and $y_0 = 4$ in (3.16) and solving for C. One finds $C = 6$, so this particular solution is

$$y_k = 6 \cdot 2^k - k - 2.$$

In the next section we show that the same methodology, with some unavoidable complications, applies to the problem of finding the general solution of the linear second-order difference equation having constant coefficients. In what follows, we shall be concentrating on the second-order equation. This is not a serious restriction in view of the fact that the theory of the nth-order ($n > 2$) linear difference equation with constant coefficients is easily obtainable as a natural extension of the theory for the second-order equation. The results for the general case of order n are summarized in Section 3.7.

PROBLEMS 3.1

1. In the general forms (3.5) through (3.7) for the linear difference equations with constant coefficients of orders 1, 2, and 3, identify the values of the coefficients a_1, a_2, a_3, and the right-hand member r_k if

(a) $y_{k+1} + 3y_k = 8$

(b) $y_{k+1} - y_k = 1$

(c) $2y_{k+1} - 5y_k = 3k + 1$

(d) $y_{k+2} - y_k = \cdot k^2$

(e) $y_{k+2} - y_{k+1} + 3y_k = 6$

(f) $2y_{k+3} - 5y_{k+2} + y_k = 2^k$

2. Write the general form of the linear difference equations with constant coefficients of orders 4 and 5. Assume in each case that the leading coefficient is 1. [*Hint:* Use (3.4) with $n = 4$ and $n = 5$.]

3. Show that each function y^* is a particular solution of the difference equation in the corresponding part of Problem 1:

(a) $y_k^* = 2$

(b) $y_k^* = k$

(c) $y_k^* = -k - 1$

(d) $y_k^* = \frac{1}{6}(k^3 - 3k^2 + 2k)$

(e) $y_k^* = 2$

(f) $y_k^* = -\frac{1}{3}2^k$

4. Using the results of Problems 3 and Theorem 3.3, find the general solutions of the difference equations in Problem 1(a), (b), (c). In each case, also find the particular solution for which $y_0 = 4$.

5. Carry out the proof of Theorem 3.1 in the two cases $n = 1$ and $n = 3$, i.e., for the first- and third-order homogeneous equations corresponding to the complete equations (3.5) and (3.7).

6. Prove the corollary to Theorem 3.1 by mathematical induction, writing out the details in the case of the difference equation of order 2.

7. In the hypothesis of Theorem 3.3, it is assumed that the difference equation is of order 1, i.e., the coefficient $a_1 \neq 0$. Where is this fact used in the proof?

3.2 FUNDAMENTAL SETS OF SOLUTIONS

We now consider the problem of solving (or, equivalently, finding the general solution of) the linear second-order difference equation with constant coefficients

$$(3.17) \qquad y_{k+2} + a_1 y_{k+1} + a_2 y_k = r_k.$$

The corresponding homogeneous equation is

$$(3.18) \qquad y_{k+2} + a_1 y_{k+1} + a_2 y_k = 0.$$

Such second-order equations have many applications in the social sciences, as we shall see in Section 3.6.

It will help in later theoretical developments if we have a concrete example before us. For this purpose, consider

$$(3.19) \qquad y_{k+2} - 5 y_{k+1} + 6 y_k = 4$$

with the homogeneous equation

$$(3.20) \qquad y_{k+2} - 5 y_{k+1} + 6 y_k = 0.$$

As in the preceding section, we attempt to find the general solution of (3.20) and a particular solution of (3.19). Then we hope to prove that the sum of these solutions is the sought-for general solution of (3.19).

It may be verified by direct substitution into (3.20) that this homogeneous difference equation has the two solutions $y^{(1)}$ and $y^{(2)}$ given by

$$(3.21) \qquad y_k^{(1)} = 2^k \qquad \text{and} \qquad y_k^{(2)} = 3^k.$$

(In the next section we develop systematic methods for finding these solutions. For the present, it is important only to check that they do indeed satisfy the equation.)

By Theorem 3.1, the function Y for which

$$(3.22) \qquad Y_k = C_1 2^k + C_2 3^k$$

is also a solution of (3.20) for any values of the constants C_1 and C_2. As a matter of fact, Y is the *general solution* of (3.20), as we now show. This means that if y is *any* solution of (3.20), then values of C_1 and C_2

can be found for which Y and y are identical. But the uniqueness theorem of Section 2.3 shows that two solutions of the linear equation of order 2 are identical if they coincide in value at two consecutive k-values, say $k = 0$ and $k = 1$. It therefore suffices to show that C_1 and C_2 may be determined so that

$$Y_0 = y_0 \quad \text{and} \quad Y_1 = y_1$$

for any numbers y_0 and y_1. From (3.22),

$$Y_0 = C_1 2^0 + C_2 3^0 = C_1 + C_2$$
$$Y_1 = C_1 2^1 + C_2 3^1 = 2C_1 + 3C_2$$

so that C_1 and C_2 must satisfy the two equations

(3.23)
$$C_1 + C_2 = y_0$$
$$2C_1 + 3C_2 = y_1.$$

This set of two simultaneous linear equations for the two unknowns C_1 and C_2 can be solved for any numbers y_0 and y_1. Multiply the first equation by 2 and subtract the result from the second equation to obtain $C_2 = y_1 - 2y_0$. Then from the first equation, $C_1 = 3y_0 - y_1$ and our proof is complete. To summarize: the solution Y may be specialized, by appropriate choices for C_1 and C_2, to agree with the values y_0 and y_1 of any solution. This makes Y and y identical, thus showing that Y is the general solution of the homogeneous difference equation.

Let us return now to the nonhomogeneous equation (3.19) which we are trying to solve. A particular solution of (3.19) is easy to find, since the constant function given by $y_k{}^* = A$ satisfies (3.19) if $A = 2$. Thus, $y_k{}^* = 2$ is the required particular solution. Again, as for the first-order equation of the preceding section, it is possible to show that the sum $y = Y + y^*$ for which

$$y_k = Y_k + y_k{}^* = C_1 2^k + C_2 3^k + 2$$

is the general solution of the complete equation (3.19). We omit this proof since it is precisely the same as that just given showing that Y is the general solution of the homogeneous equation.

Lest the reader think that we can start with *any* solutions $y^{(1)}$ and $y^{(2)}$ in place of those in (3.21), it is well to outline the consequences in the above proof if we take

$$y_k^{(1)} = 2^k \quad \text{and} \quad y_k^{(2)} = 5 \cdot 2^k$$

both of which are certainly solutions of the homogeneous equation (3.20). Then

$$Y_k = C_1 2^k + C_2 5 \cdot 2^k$$

is again a solution, but this time *not* the general solution. For if we proceed as before, the requirements $Y_0 = y_0$ and $Y_1 = y_1$ lead to the equations

(3.24)
$$C_1 + 5C_2 = y_0$$
$$2C_1 + 10C_2 = y_1.$$

But *this* set of two simultaneous linear equations for the unknowns C_1 and C_2 does not have a solution for all pairs of initial values y_0 and y_1. In fact, if $y_1 = 2y_0$, then the second equation is identical to the first after division by the factor 2. This means that we actually have just one equation in two unknowns and there are infinitely many pairs of values of C_1 and C_2 satisfying such an equation. On the other hand, if $y_1 \neq 2y_0$, then the two equations in (3.24) are incompatible and there is no solution.

This illustrative example raises a number of questions for the general theory of equation (3.17):

I. What condition(s) must be met by two particular solutions $y^{(1)}$ and $y^{(2)}$ of the homogeneous equation (3.18) in order that the function Y given by

$$Y_k = C_1 y_k^{(1)} + C_2 y_k^{(2)},$$

with C_1 and C_2 arbitrary constants, be the general solution of (3.18)?

II. If y^* is a particular solution of the complete equation (3.17) and Y is the general solution of the homogeneous equation (3.18), is $Y + y^*$ the general solution of (3.17)?

III. Are there systematic methods for obtaining the particular solutions $y^{(1)}$, $y^{(2)}$, and y^* required in I and II?

We answer the first two questions in this section and Question III (in the affirmative) in Sections 3.3 and 3.4.

We need to know a result from algebra concerning systems of simultaneous linear equations. The system of two equations in the two unknown x and y

(3.25) $$ax + by = e \qquad cx + dy = f,$$

where a, b, c, d, e, and f are constants, is said to have a solution (x_0, y_0) if both equations are true when $x = x_0$ and $y = y_0$. The fundamental theorem for our purpose is as follows:

THEOREM 3.4. *The system of simultaneous equations* (3.25) *has a unique solution if and only if* $ad - bc \neq 0$.

The number $ad - bc$ is often written as a *determinant* in the form

$$\begin{vmatrix} a & b \\ c & d \end{vmatrix} = ad - bc.$$

This number is said to be the determinant formed by the coefficients of x and y in the system (3.25). The reader may check that the determinant of the system (3.23), for which we found a unique solution, is

$$\begin{vmatrix} 1 & 1 \\ 2 & 3 \end{vmatrix} = 1 \cdot 3 - 1 \cdot 2 = 1 \neq 0,$$

whereas that of (3.24), for which no unique solution exists, is

$$\begin{vmatrix} 1 & 5 \\ 2 & 10 \end{vmatrix} = 1 \cdot 10 - 5 \cdot 2 = 0.$$

A proof of Theorem 3.4 together with additional facts about determinants may be found in most algebra textbooks.

We are now able to answer Question I.

THEOREM 3.5. *Let $y^{(1)}$ and $y^{(2)}$ be two solutions of the homogeneous difference equation* (3.18) *and let*

$$Y = C_1 y^{(1)} + C_2 y^{(2)},$$

where C_1 and C_2 are arbitrary constants. If

$$(3.26) \qquad \begin{vmatrix} y_0^{(1)} & y_0^{(2)} \\ y_1^{(1)} & y_1^{(2)} \end{vmatrix} = y_0^{(1)} y_1^{(2)} - y_0^{(2)} y_1^{(1)} \neq 0,$$

then Y is the general solution of (3.18).

PROOF. The proof is similar to that already given for the difference equation (3.20). By Theorem 3.1, Y is *a* solution of (3.18). We need only show that if y is *any* solution of (3.18), C_1 and C_2 may be determined so that Y and y are identical. By the uniqueness Theorem 2.1, it suffices to show that Y and y are equal at $k = 0$ and $k = 1$. That is, we must determine C_1 and C_2 so that $Y_0 = y_0$ and $Y_1 = y_1$ for any choice of y_0 and y_1. But

$$Y_0 = C_1 y_0^{(1)} + C_2 y_0^{(2)}$$

and

$$Y_1 = C_1 y_1^{(1)} + C_2 y_1^{(2)}$$

so C_1 and C_2 must satisfy the equations

$$y_0^{(1)} C_1 + y_0^{(2)} C_2 = y_0$$
$$y_1^{(1)} C_1 + y_1^{(2)} C_2 = y_1.$$

This is a set of two simultaneous linear equations for the unknowns C_1 and C_2. By hypothesis, the determinant formed by the coefficients of C_1 and C_2 is different from zero. By Theorem 3.4, we can solve to find a unique pair of values of C_1 and C_2 for each choice of y_0 and y_1. This concludes the proof.

DEFINITION 3.1. *Two solutions, $y^{(1)}$ and $y^{(2)}$, of* (3.18) *which satisfy condition* (3.26) *are said to form a fundamental set* (*or fundamental system*) *of solutions of* (3.18).

This terminology may be used to restate Theorem 3.5: *The general solution of* (3.18) *is given by*

$$Y = C_1 y^{(1)} + C_2 y^{(2)},$$

where $y^{(1)}$ and $y^{(2)}$ form a fundamental set of solutions and C_1 and C_2 are constants.

For the complete equation (3.17) we have the following result, which answers Question II.

THEOREM 3.6. *If y^* is a particular solution of* (3.17) *and $y^{(1)}$ and $y^{(2)}$ form a fundamental set of solutions of* (3.18), *then the general solution of* (3.17) *is given by*

$$y = C_1 y^{(1)} + C_2 y^{(2)} + y^*,$$

where C_1 and C_2 are constants.

PROOF. That y is a solution of (3.17) follows from Theorem 3.2. That it is the general solution is proved precisely as the corresponding result for Y in Theorem 3.5. The only change is that, in the present case, C_1 and C_2 must exist which satisfy the simultaneous equations

$$y_0^{(1)} C_1 + y_0^{(2)} C_2 = y_0 - y_0^*$$
$$y_1^{(1)} C_1 + y_1^{(2)} C_2 = y_1 - y_1^*.$$

As before, the hypothesis that $y^{(1)}$ and $y^{(2)}$ form a fundamental set ensures the existence of C_1 and C_2 for any choice of y_0 and y_1.

Example

The difference equation

(3.27) $$y_{k+2} - 3y_{k+1} + 2y_k = -1$$

has the particular solution given by $y_k^* = k$, as may be verified by direct substitution. Similarly, it may be shown that the homogeneous equation

(3.28) $$y_{k+2} - 3y_{k+1} + 2y_k = 0$$

has the two solutions $y^{(1)}$ and $y^{(2)}$ given by

$$y_k^{(1)} = 1 \quad \text{and} \quad y_k^{(2)} = 2^k.$$

(Again we remind the reader that systematic methods for obtaining these solutions are yet to be considered. The present purpose is to show that they are worth having.)

These two solutions form a fundamental set since

$$\begin{vmatrix} y_0^{(1)} & y_0^{(2)} \\ y_1^{(1)} & y_1^{(2)} \end{vmatrix} = \begin{vmatrix} 1 & 1 \\ 1 & 2 \end{vmatrix} = 2 - 1 = 1 \neq 0.$$

Hence, by Theorems 3.5 and 3.6,

$$Y_k = C_1 + C_2 2^k$$

gives the general solution of the homogeneous equation (3.28) and

(3.29) $$y_k = C_1 + C_2 2^k + k$$

the general solution of (3.27).

If we want a solution of (3.27) for which

(3.30) $$y_0 = 0 \quad \text{and} \quad y_1 = 3,$$

we determine the values of C_1 and C_2 as follows. From (3.29), putting $k = 0$ and $k = 1$ and using the initial conditions (3.30), we obtain

$$0 = C_1 + C_2$$
$$3 = C_1 + 2C_2 + 1.$$

Therefore we must solve the system of equations

$$C_1 + C_2 = 0$$
$$C_1 + 2C_2 = 2.$$

One obtains the solution $C_1 = -2$, $C_2 = 2$. Thus the solution given by

$$y_k = -2 + 2^{k+1} + k$$

satisfies *both* the difference equation (3.27) and the initial conditions (3.30).

It may similarly be shown that if the initial conditions are

$$y_0 = -3 \quad \text{and} \quad y_1 = 5,$$

we must put $C_1 = -10$ and $C_2 = 7$ to obtain the solution

$$y_k = -10 + 7 \cdot 2^k + k.$$

In general, each set of initial conditions prescribing any two consecutive y-values (not necessarily y_0 and y_1) will enable us to determine the values of C_1 and C_2 for which the solution (3.29) satisfies these conditions.

PROBLEMS 3.2

1. Show that the functions $y^{(1)}$ and $y^{(2)}$ given by

$$y_k^{(1)} = 1 \quad \text{and} \quad y_k^{(2)} = (-1)^k$$

are solutions of the difference equation

$$y_{k+2} - y_k = 0$$

and that they form a fundamental set. Find the general solution of the difference equation and a particular solution for which $y_0 = 0$ and $y_1 = 2$.

2. Find the particular solution of the difference equation in Problem 1 which satisfies the initial conditions $y_1 = 2$ and $y_2 = 0$. Note that this is the same solution as that obtained in Problem 1. Explain.

3. Use the result of 3(d) of Problems 3.1 and Problem 1 to find the general solution of the difference equation

$$y_{k+2} - y_k = k^2.$$

Find a particular solution of this equation for which $y_0 = 0$ and $y_1 = 2$.

4. In each of the following parts, a difference equation and three functions, $y^{(1)}$, $y^{(2)}$, and y^*, are given. Show in each case that (a) $y^{(1)}$ and $y^{(2)}$ are solutions of the corresponding homogeneous equation; (b) $y^{(1)}$ and $y^{(2)}$ form a fundamental set of solutions; (c) y^* is a solution of the complete equation. Then (d) use Theorem 3.6 to write the general solution of the complete equation, and (e) find the particular solution satisfying the initial conditions $y_0 = 0$ and $y_1 = 1$.

(i) $y_{k+2} - 4y_{k+1} + 4y_k = 1$ $y_k^{(1)} = 2^k$ $y_k^{(2)} = k2^k$ $y_k^* = 1$

(ii) $y_{k+2} - 7y_{k+1} + 12y_k = 2$ $y_k^{(1)} = 3^k$ $y_k^{(2)} = 4^k$ $y_k^* = \frac{1}{3}$

(iii) $y_{k+2} - 4y_k = 9$ $y_k^{(1)} = 2^k$ $y_k^{(2)} = (-2)^k$ $y_k^* = -3$

(iv) $2y_{k+2} + 3y_{k+1} - 2y_k = 3k + 1$ $y_k^{(1)} = (\frac{1}{2})^k$ $y_k^{(2)} = (-2)^k$ $y_k^* = k - 2$

(v) $8y_{k+2} - 6y_{k+1} + y_k = 2^k$ $y_k^{(1)} = (\frac{1}{2})^k$ $y_k^{(2)} = (\frac{1}{4})^k$ $y_k^* = \frac{1}{21} 2^k$

5. Let $y^{(1)}$ and $y^{(2)}$ be any two solutions of (3.18) and let W be the function whose value at k is given by

$$W_k = \begin{vmatrix} y_k^{(1)} & y_k^{(2)} \\ y_{k+1}^{(1)} & y_{k+1}^{(2)} \end{vmatrix} = y_k^{(1)} y_{k+1}^{(2)} - y_k^{(2)} y_{k+1}^{(1)}.$$

(Note that $W_0 \neq 0$ is the condition for $y^{(1)}$ and $y^{(2)}$ to form a fundamental set.) Write W_{k+1} and use the fact that $y^{(1)}$ and $y^{(2)}$ satisfy (3.18) to show that $W_{k+1} = a_2 W_k$. But $a_2 \neq 0$ (why?). Conclude that either W is identically zero, i.e., $W_k = 0$ for all k, or W is never zero.

3.3 GENERAL SOLUTION OF THE HOMOGENEOUS EQUATION

In view of Theorem 3.5 the problem of finding the general solution of the homogeneous equation

$$(3.31) \qquad\qquad y_{k+2} + a_1 y_{k+1} + a_2 y_k = 0$$

reduces to the problem of finding two solutions which form a fundamental set, i.e., for which (3.26) is satisfied. This problem is completely solved in this section.

It is easy to find solutions of the difference equation for which

$$(3.32) \qquad\qquad y_k = m^k$$

where m is some suitably chosen constant different from zero. (We disallow $m = 0$ because it makes y identically zero, a solution which cannot be one of a fundamental set.) If this trial solution is substituted in (3.31), we obtain, after division by the common factor m^k,

$$(3.33) \qquad\qquad m^2 + a_1 m + a_2 = 0.$$

This quadratic equation is called the *auxiliary equation* (or characteristic equation) of the difference equation (3.31). We have shown that if m is a number satisfying the auxiliary equation (m is then called a *root* of the equation), (3.32) is a solution of the difference equation (3.31).

The auxiliary equation is a quadratic algebraic equation and therefore has two nonzero roots, say m_1 and m_2. [Zero roots are excluded since $a_2 \neq 0$, a requirement which is needed for (3.31) to be of second order.] To these roots there correspond the solutions $y^{(1)}$ and $y^{(2)}$ given by

$$(3.34) \qquad y_k^{(1)} = m_1{}^k \qquad \text{and} \qquad y_k^{(2)} = m_2{}^k.$$

It is now most convenient to consider three cases: (1) the roots m_1 and m_2 are real numbers and unequal, (2) the two roots are real and equal, and (3) the two roots are complex numbers. These exhaust the possibilities for a quadratic equation since it is impossible to have one real and one complex root. The following examples illustrate these three cases.

Examples

(a) The homogeneous difference equation

$$(3.35) \qquad y_{k+2} - 3y_{k+1} + 2y_k = 0$$

has the auxiliary equation

$$m^2 - 3m + 2 = 0.$$

This equation may be factored since

$$(m^2 - 3m + 2) = (m - 1)(m - 2).$$

Hence the two roots, $m_1 = 1$ and $m_2 = 2$, are real and unequal.

(b) The difference equation

$$(3.36) \qquad y_{k+2} - 2y_{k+1} + y_k = 0$$

has the auxiliary equation

$$m^2 - 2m + 1 = 0.$$

Thus

$$(m - 1)(m - 1) = 0$$

and the two roots, $m_1 = 1$ and $m_2 = 1$, are real and equal.

(c) The difference equation

$$(3.37) \qquad y_{k+2} + y_k = 0$$

has the auxiliary equation

$$m^2 + 1 = 0 \qquad \text{or} \qquad m^2 = -1.$$

This quadratic equation has the two roots $m_1 = i$ and $m_2 = -i$, where $i = \sqrt{-1}$. These roots, involving square roots of a negative number, are complex numbers.

CASE 1. ROOTS REAL AND UNEQUAL

In this case, the solutions in (3.34) form a fundamental set. To prove this, we calculate the determinant as given in Theorem 3.5. We obtain

$$(3.38) \qquad \begin{vmatrix} y_0^{(1)} & y_0^{(2)} \\ y_1^{(1)} & y_1^{(2)} \end{vmatrix} = \begin{vmatrix} 1 & 1 \\ m_1 & m_2 \end{vmatrix} = m_2 - m_1$$

which is different from zero since $m_1 \neq m_2$. Hence, by Theorem 3.5, the general solution of the homogeneous difference equation (3.31) is given by

$$(3.39) \qquad Y_k = C_1 m_1{}^k + C_2 m_2{}^k.$$

Examples

(a) The difference equation (3.35) was found to have an auxiliary equation with real, unequal roots $m_1 = 1$ and $m_2 = 2$. Hence, the general solution of (3.35) is given by

$$Y_k = C_1 1^k + C_2 2^k = C_1 + C_2 2^k.$$

(b) The difference equation

$$(3.40) \qquad y_{k+2} + 3y_{k+1} + y_k = 0$$

has the auxiliary equation

$$m^2 + 3m + 1 = 0.$$

The roots of this equation (as of *any* quadratic equation) may be calculated using the quadratic formula.[1] These roots are

$$\frac{-3 \pm \sqrt{9-4}}{2} = \frac{-3 \pm \sqrt{5}}{2}.$$

These are real and unequal roots, so the general solution of (3.40) is given by

$$Y_k = C_1 \left(\frac{-3 + \sqrt{5}}{2} \right)^k + C_2 \left(\frac{-3 - \sqrt{5}}{2} \right)^k$$

CASE 2. ROOTS REAL AND EQUAL

Now the solutions $y^{(1)}$ and $y^{(2)}$ in (3.34) no longer form a fundamental set since the determinant (3.38) is zero. We keep $y^{(1)}$, but must find a new function $y^{(2)}$ which together with $y^{(1)}$ will form a fundamental set. Such a solution is given by

$$(3.41) \qquad y_k^{(2)} = k m_1{}^k.$$

---

[1] The roots of the quadratic equation $ax^2 + bx + c = 0$ are

$$\frac{-b \pm \sqrt{b^2 - 4ac}}{2a},$$

one root being obtained using the plus sign, the other using the minus sign before the square root. By adding and multiplying these roots, one may show that their sum is $-b/a$ and their product is c/a.

[A systematic, but somewhat involved, procedure exists for *deriving* this solution, as well as those in (3.34), rather than "guessing" them as we have done. See 4 of Problems 3.3.]

To prove that if $m_1 = m_2$, then (3.41) actually is a solution, we substitute $y^{(2)}$ for y in the difference equation (3.31) and check that the equation is satisfied:

$$y^{(2)}_{k+2} + a_1 y^{(2)}_{k+1} + a_2 y^{(2)}_k = (k+2)m_1^{k+2} + a_1(k+1)m_1^{k+1} + a_2 k m_1^k$$

(3.42)
$$= k m_1^k (m_1^2 + a_1 m_1 + a_2) + m_1^{k+1}(2m_1 + a_1).$$

But the terms in the first parenthesis add to 0 since m_1 is a root of the auxiliary equation. Furthermore, the sum of the roots of the auxiliary equation (see footnote 1) is $-a_1$. But if $m_1 = m_2$, this sum is $2m_1$. Hence $2m_1 + a_1 = 0$. This makes (3.42) identically zero, so (3.41) is indeed a solution of the difference equation.

Moreover, the two solutions

$$y^{(1)}_k = m_1^k \qquad \text{and} \qquad y^{(2)}_k = k m_1^k$$

form a fundamental set. For

$$\begin{vmatrix} y^{(1)}_0 & y^{(2)}_0 \\ y^{(1)}_1 & y^{(2)}_1 \end{vmatrix} = \begin{vmatrix} 1 & 0 \\ m_1 & m_1 \end{vmatrix} = m_1$$

which is different from zero since, as we have already noted, no root of the auxiliary equation is zero.

Thus, when the two roots m_1 and m_2 are equal, the general solution of (3.31) is given by

$$Y_k = C_1 m_1^k + C_2 k m_1^k$$

or

(3.43)
$$Y_k = (C_1 + C_2 k)m_1^k.$$

Examples

(a) The difference equation (3.36) was found to have an auxiliary equation with the two equal roots $m_1 = m_2 = 1$. Hence the general solution of (3.36) is given by

$$Y_k = C_1 + C_2 k.$$

(b) The difference equation

(3.44)
$$4y_{k+2} + 4y_{k+1} + y_k = 0$$

has the auxiliary equation

$$4m^2 + 4m + 1 = 0.$$

[Technically, we should first put (3.44) into standard form and then write $m^2 + m + \frac{1}{4} = 0$ as the auxiliary equation. The roots are the same, of course.] This quadratic equation factors into

$$(2m + 1)^2 = 0$$

so that $m_1 = m_2 = -\frac{1}{2}$. Hence the general solution of (3.44) is given by

$$y_k = (C_1 + C_2 k)(-\tfrac{1}{2})^k.$$

CASE 3.   COMPLEX ROOTS

Before considering the case of complex roots of the auxiliary equation, we review some material on complex numbers. Any complex number can be written in the form $a + bi$, where a and b are real numbers and i is the "imaginary" unit for which $i^2 = -1$. The numbers $a + bi$ and $a - bi$ are said to be *complex conjugates* (or a conjugate pair). The complex number $a + bi$ can be represented graphically, using a rectangular coordinate system as in Figure 3.1, by the point P with coordinates (a, b). However, point P can also be identified by its polar coordinates, i.e., by specifying the length r of the line segment from the origin to P and the angle θ that this segment makes with the positive x-axis. Since $a = r \cos \theta$ and $b = r \sin \theta$, it follows that

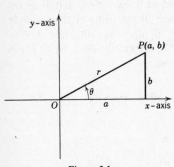

$$a + bi = r(\cos \theta + i \sin \theta).$$

Figure 3.1

This latter form is called the *polar* or *trigonometric* form of the complex number $a + bi$.

The angle θ is determined only to within integral multiples of 360°. But, if counterclockwise (from the positive x-axis) rotations about O are represented by positive angles and clockwise rotations by negative angles, we account for all possible positions of point P by allowing θ to vary only in the interval $-\pi < \theta \leq \pi$. (We use a radian angle measure: π radians $= 180°$.) Now the identity $\sin^2 \theta + \cos^2 \theta = 1$ shows that $a^2 + b^2 = r^2$ or

(3.45) $$r = \sqrt{a^2 + b^2}.$$

Hence, θ may be taken as the unique angle such that

(3.46) $$\cos \theta = \frac{a}{\sqrt{a^2 + b^2}} \qquad \sin \theta = \frac{b}{\sqrt{a^2 + b^2}} \qquad -\pi < \theta \leq \pi.$$

Formulas (3.45) and (3.46) are used to find the polar form of a complex number. The positive length r is called the *modulus* of the number and the angle θ, its *amplitude*. It is advisable to plot the complex number (as in Figure 3.1) as an aid in determining its modulus and amplitude.

Arithmetic operations are defined for complex numbers in such a way that the ordinary algebraic rules can be used with the proviso that any resulting expression be simplified by using the equations $i^2 = -1$, $i^3 = -i$, $i^4 = 1, \cdots$. In polar form, the product of two complex numbers is a complex number whose modulus is the product of their moduli and whose amplitude is the sum of their amplitudes. For by direct multiplication, using $i^2 = -1$, we find that

$$r_1(\cos \theta_1 + i \sin \theta_1) \cdot r_2(\cos \theta_2 + i \sin \theta_2)$$
$$= r_1 r_2[(\cos \theta_1 \cos \theta_2 - \sin \theta_1 \sin \theta_2) + i(\sin \theta_1 \cos \theta_2 + \cos \theta_1 \sin \theta_2)]$$
$$= r_1 r_2[\cos (\theta_1 + \theta_2) + i \sin (\theta_1 + \theta_2)],$$

where we have used the standard trigonometric formulas for the sine and cosine of the sum of two angles (cf. 8 of Problems 1.6). By repeated applications of this multiplication rule, we find that *if n is any positive integer,*

$$[r(\cos \theta + i \sin \theta)]^n = r^n(\cos n\theta + i \sin n\theta),$$

a result known as *de Moivre's theorem.* A formal proof by mathematical induction, together with an expanded treatment of all of these topics, can be found in many textbooks.[2]

Example

To write the complex number $1 + i$ in polar form, we calculate $r = \sqrt{1^2 + 1^2}$ $= \sqrt{2}$, $\cos \theta = 1/\sqrt{2}$, and $\sin \theta = 1/\sqrt{2}$. Hence [this is obvious from a plot of $1 + i$ since the line segment from the origin to $(1, 1)$ bisects the first quadrant], $\theta = \pi/4$ or $45°$. Therefore

$$1 + i = \sqrt{2} \left(\cos \frac{\pi}{4} + i \sin \frac{\pi}{4} \right).$$

By de Moivre's theorem,

$$(1 + i)^5 = (\sqrt{2})^5 \left(\cos \frac{5\pi}{4} + i \sin \frac{5\pi}{4} \right)$$

$$= 4\sqrt{2} \left(\cos \frac{-3\pi}{4} + i \sin \frac{-3\pi}{4} \right)$$

$$= 4\sqrt{2} \left(\frac{-1}{\sqrt{2}} + i \frac{-1}{\sqrt{2}} \right)$$

$$= -4 - 4i,$$

a result that can be checked by expanding $(1 + i)^5$ using the binomial theorem.

For our present purposes, it is noteworthy that *complex roots of a quadratic equation always occur in conjugate pairs.* Therefore, if m_1 and m_2 are complex roots of the auxiliary equation, then $m_1 \neq m_2$, and the calculation in (3.38) shows that the two solutions $y_k^{(1)} = m_1^k$ and $y_k^{(2)} = m_2^k$ form a fundamental set. The general solution is given in (3.39). The only difficulty with this general solution is that

$$Y_k = C_1 m_1^k + C_2 m_2^k$$

[2] E.g., E. A. Cameron and E. T. Browne, *College Algebra*, rev. ed., Henry Holt, New York, 1956, Chap. 15.

may be a complex number if m_1 and m_2 are themselves complex. We require solutions which have real number values for all k-values for which they are defined. It is possible to show that if C_1 and C_2 are complex conjugates, Y_k is always a real number. To prove this, we write all complex numbers in *polar form*.

The roots of the auxiliary equation are complex conjugates and hence have polar form

$$(3.47) \qquad m_1 = r(\cos \theta + i \sin \theta) \qquad m_2 = r(\cos \theta - i \sin \theta).$$

We are assuming C_1 and C_2 are complex conjugates:

$$C_1 = a(\cos B + i \sin B) \qquad C_2 = a(\cos B - i \sin B).$$

By de Moivre's theorem,

$$m_1{}^k = r^k(\cos k\theta + i \sin k\theta) \qquad m_2{}^k = r^k(\cos k\theta - i \sin k\theta).$$

Therefore, using the rules for multiplying complex numbers, we obtain the following simplification for Y_k:

$$
\begin{aligned}
Y_k = C_1 m_1{}^k + C_2 m_2{}^k &= ar^k[\cos (k\theta + B) + i \sin (k\theta + B)] \\
&\quad + ar^k[\cos (k\theta + B) - i \sin (k\theta + B)] \\
(3.48) \qquad\qquad &= 2ar^k \cos (k\theta + B).
\end{aligned}
$$

And this is a real number, as claimed.

The numbers r and θ are determined from (3.47) by writing the roots of the auxiliary equation in polar form, using (3.45) and (3.46) to find r and θ. The constants a and B in (3.48) take the place of C_1 and C_2. If we denote the constant $2a$ by A, then the general solution of the homogeneous difference equation when the auxiliary equation has complex roots can be written in the form

$$(3.49) \qquad\qquad Y_k = Ar^k \cos (k\theta + B),$$

where A and B are arbitrary constants.

Examples

(*a*) The difference equation (3.37) was found to have an auxiliary equation $m^2 + 1 = 0$ with roots $m_1 = i$ and $m_2 = -i$. To write m_1 in polar form we use (3.45) and (3.46). (Note that $i = 0 + 1i$.) We find $r = 1$, and θ must be determined so that $\cos \theta = 0$ and $\sin \theta = 1$. Hence

$$\theta = \frac{\pi}{2} \qquad \text{and} \qquad m_1 = i = \cos \frac{\pi}{2} + i \sin \frac{\pi}{2}.$$

Therefore, from (3.49), the general solution is given by

$$(3.50) \qquad\qquad Y_k = A \cos \left(\frac{k\pi}{2} + B\right).$$

If we seek a particular solution of (3.37) satisfying the initial conditions $y_0 = 0$ and $y_1 = 1$, we find from (3.50) that A and B must satisfy the equations

$$0 = A \cos B \qquad 1 = A \cos\left(\frac{\pi}{2} + B\right).$$

The first of these is satisfied by $B = \pi/2$. Then the second becomes $1 = A \cos \pi$, which makes $A = -1$, since $\cos \pi = -1$. We find the required particular solution of (3.37) by putting these values of A and B in (3.50):

$$y_k = -\cos\left(\frac{k\pi}{2} + \frac{\pi}{2}\right).$$

But for any θ,

$$\cos\left(\theta + \frac{\pi}{2}\right) = -\sin\theta.$$

Hence

$$y_k = \sin\frac{k\pi}{2}$$

yields that solution of (3.37) for which $y_0 = 0$ and $y_1 = 1$.

(b) The difference equation

(3.51)
$$y_{k+2} - 2y_{k+1} + 2y_k = 0$$

has an auxiliary equation $m^2 - 2m + 2 = 0$. From the quadratic formula, we find the roots $m_1 = 1 + i$ and $m_2 = 1 - i$. In polar form,

$$m_1 = 1 + i = \sqrt{2}\left(\cos\frac{\pi}{4} + i\sin\frac{\pi}{4}\right).$$

Therefore, the general solution of (3.51) is given by

$$Y_k = A(\sqrt{2})^k \cos\left(\frac{k\pi}{4} + B\right).$$

The following theorem summarizes our results.

THEOREM 3.7. *If*

(3.52)
$$y_{k+2} + a_1 y_{k+1} + a_2 y_k = 0,$$

where a_1 and a_2 are constants and $a_2 \neq 0$, and if m_1 and m_2 are the two roots of the auxiliary equation

(3.53)
$$m^2 + a_1 m + a_2 = 0,$$

the general solution of the difference equation (3.52) is given by

(3.54)
$$Y_k = C_1 m_1{}^k + C_2 m_2{}^k$$

if m_1 and m_2 are real and unequal; by

(3.55)
$$Y_k = (C_1 + C_2 k)m_1{}^k$$

if m_2 is equal to m_1; and by

(3.56) $$Y_k = Ar^k \cos (k\theta + B)$$

if m_1 and m_2 are complex conjugates with polar forms

(3.57) $$r(\cos \theta \pm i \sin \theta).$$

PROBLEMS 3.3

1. Find the general solution of each of the following homogeneous difference equations:

(a) $y_{k+2} - y_k = 0$
(b) $2y_{k+2} - 5y_{k+1} + 2y_k = 0$
(c) $y_{k+2} + 2y_{k+1} + y_k = 0$
(d) $9y_{k+2} - 6y_{k+1} + y_k = 0$
(e) $3y_{k+2} - 6y_{k+1} + 4y_k = 0$
(f) $y_{k+2} + 6y_{k+1} + 25y_k = 0$

2. For each of the equations in Problem 1, find a particular solution satisfying the initial conditions $y_0 = 0$ and $y_1 = 1$.

3. For each of the particular solutions obtained in Problem 2, find the first seven terms of the solution sequence $\{y_k\}$.   Conjecture as to the type of sequence obtained.   (Review Section 2.5 first.)

4. Write the difference equation (3.52) in operator notation as

$$(E^2 + a_1 E + a_2)y_k = 0,$$

where E is the displacement operator of the calculus of finite differences (see Section 1.3).   If m_1 and m_2 are roots of the auxiliary equation (3.53), show that (see Section 1.5)

$$(E^2 + a_1E + a_2) \equiv (E - m_1)(E - m_2).$$

Hence the difference equation becomes

(3.58) $$(E - m_1)(E - m_2)y_k = 0.$$

Let z be a new function for which

(3.59) $$(E - m_2)y_k = z_k.$$

Then (3.58) becomes

$$(E - m_1)z_k = 0.$$

Solve this first-order difference equation and thus show that

$$z_k = m_1{}^k$$

is a solution.   With this z, show that (3.59) may be written in the form

$$y_{k+1} - m_2 y_k = m_1{}^k.$$

Finally show (cf. 8 of Problems 2.4) that the solution of this equation is of the form

$$y_k = C_1 m_1{}^k + C_2 m_2{}^k$$

if $m_1 \neq m_2$, and

$$y_k = C_1 m_1{}^k + C_2 k m_1{}^k$$

if $m_1 = m_2$.

3.4 PARTICULAR SOLUTIONS OF THE COMPLETE EQUATION

We turn now to methods for finding solutions of the complete equation

$$(3.60) \qquad y_{k+2} + a_1 y_{k+1} + a_2 y_k = r_k.$$

Having already found the general solution of the corresponding homogeneous equation, if we add to it *any* solution of (3.60), the sum, according to Theorem 3.6, will be the general solution of the complete equation.

A number of special techniques exist for finding particular solutions of (3.60). The most useful of these is the *method of undetermined coefficients*. Some examples will illustrate this method and serve to clarify the general discussion which follows.

Examples

(a) Consider the equation

$$(3.61) \qquad y_{k+2} - 3y_{k+1} + 2y_k = 3^k$$

and let us try to find a solution of the form

$$y_k{}^* = A3^k$$

where the constant coefficient A is as yet undetermined. We seek a value of A, if one exists, for which y^* will be a solution of (3.61). Now

$$\begin{aligned} y_{k+2}^* - 3y_{k+1}^* + 2y_k{}^* &= A3^{k+2} - 3A3^{k+1} + 2A3^k \\ &= A3^k(9 - 9 + 2) \\ &= 2A3^k. \end{aligned}$$

In order that y^* satisfy (3.61), we must choose $A = \frac{1}{2}$. Then $y_k{}^* = \frac{1}{2}3^k$ is the desired particular solution and (since the auxiliary equation $m^2 - 3m + 2 = 0$ has the roots $m_1 = 1$ and $m_2 = 2$) the general solution of (3.61) is given by

$$y_k = C_1 + C_2 2^k + \tfrac{1}{2}3^k,$$

where C_1 and C_2 are arbitrary constants.

(b) We now attempt to find a particular solution of the more general equation

$$(3.62) \qquad y_{k+2} - 3y_{k+1} + 2y_k = a^k,$$

where a is some constant. [Equation (3.61) is the special case $a = 3$.] As in Example (a), we attempt a trial solution of the form

$$(3.63) \qquad y_k{}^* = Aa^k$$

and obtain

$$y^*_{k+2} - 3y^*_{k+1} + 2y_k^* = Aa^{k+2} - 3Aa^{k+1} + 2Aa^k$$
$$= Aa^k(a^2 - 3a + 2)$$
$$= Aa^k(a - 1)(a - 2).$$

Hence we require that

$$A(a - 1)(a - 2) = 1,$$

an equation from which A may be found provided that a is not equal to 1 or 2, i.e., *provided that a is not a root of the auxiliary equation.*

Thus, if a is not equal to 1 or 2,

$$y_k^* = \frac{1}{(a - 1)(a - 2)} a^k$$

yields a particular solution of (3.62).

Separate considerations are required if $a = 1$ or if $a = 2$, for in these two cases there is no value of the coefficient A for which the trial function in (3.63) becomes a solution of the nonhomogeneous difference equation (3.62). In fact, for every value of A, (3.63) is a solution of the corresponding homogeneous equation if a is a root of the auxiliary equation, i.e., if $a = 1$ or $a = 2$. We here anticipate the theory to be outlined below which tells us, in these circumstances, first to modify the trial solution (3.63) by multiplying by k and then to determine the coefficient A so that the new function given by $y_k^* = Aka^k$ becomes a solution of (3.62). We perform the calculations in the case $a = 1$, i.e., when (3.62) becomes

$$y_{k+2} - 3y_{k+1} + 2y_k = 1$$

and the trial function is $y_k^* = Ak$. Now

$$y^*_{k+2} - 3y^*_{k+1} + 2y_k^* = A(k + 2) - 3A(k + 1) + 2Ak = -A$$

so the choice $A = -1$ yields a particular solution given by $y_k^* = -k$. Thus, if $a = 1$, the general solution of (3.62) is

$$y_k = C_1 + C_2 2^k - k.$$

The reader may similarly show that when $a = 2$, the trial function becomes $y_k^* = Ak2^k$ and the coefficient A must be equal to $\frac{1}{2}$ if y^* is to be a solution of (3.62). This difference equation,

$$y_{k+2} - 3y_{k+1} + 2y_k = 2^k,$$

is thereby found to have the general solution given by

$$y_k = C_1 + C_2 2^k + k2^{k-1}.$$

(*c*) To find a particular solution of

(3.64) $$8y_{k+2} - 6y_{k+1} + y_k = 5 \sin \frac{k\pi}{2}$$

we try a solution of the form

$$y_k^* = A \sin \frac{k\pi}{2} + B \cos \frac{k\pi}{2},$$

where A and B are constant coefficients to be determined.   Now

$$8y^*_{k+2} - 6y^*_{k+1} + y_k{}^* = 8 \left[A \sin \frac{(k+2)\pi}{2} + B \cos \frac{(k+2)\pi}{2} \right]$$

$$- 6 \left[A \sin \frac{(k+1)\pi}{2} + B \cos \frac{(k+1)\pi}{2} \right] + \left[A \sin \frac{k\pi}{2} + B \cos \frac{k\pi}{2} \right].$$

Using the trigonometric identities

$$\sin \frac{(k+2)\pi}{2} = \sin \left(\frac{k\pi}{2} + \pi \right) = -\sin \frac{k\pi}{2}$$

$$\cos \frac{(k+2)\pi}{2} = \cos \left(\frac{k\pi}{2} + \pi \right) = -\cos \frac{k\pi}{2}$$

$$\sin \frac{(k+1)\pi}{2} = \sin \left(\frac{k\pi}{2} + \frac{\pi}{2} \right) = \cos \frac{k\pi}{2}$$

$$\cos \frac{(k+1)\pi}{2} = \cos \left(\frac{k\pi}{2} + \frac{\pi}{2} \right) = -\sin \frac{k\pi}{2},$$

this simplifies to

$$8y^*_{k+2} - 6y^*_{k+1} + y_k{}^* = (-7A + 6B) \sin \frac{k\pi}{2} + (-6A - 7B) \cos \frac{k\pi}{2}.$$

If y^* is to satisfy (3.64), A and B must satisfy the equations

$$-7A + 6B = 5 \qquad -6A - 7B = 0.$$

Multiplying the first of these equations by 7 and the second by 6 and adding, we obtain $-85A = 35$ or $A = -\frac{7}{17}$.   Then we find $B = \frac{6}{17}$.   Hence

$$y_k{}^* = \tfrac{1}{17} \left(-7 \sin \frac{k\pi}{2} + 6 \cos \frac{k\pi}{2} \right)$$

is a particular solution of (3.64).

From these examples we learn that the form of a particular solution may often be inferred from the form of the function r in the complete equation (3.60).   The method of undetermined coefficients can be applied successfully when the function r is a linear combination of sums and products of functions with the following values: a^k, $\sin bk$, $\cos bk$, and k^n, where a and b are any constants and n is any nonnegative integer.[3]

[3] If the *family* of a function f is defined as the set of all functions of which f and Ef, E^2f, E^3f, $\cdots$ are linear combinations, then only these functions have *finite* families. For example, the family of the function f given by $f_k = a^k$ consists of the single function f (since Ef, E^2f, E^3f, $\cdots$ are all just constant multiples of f) whereas the family of the function given by $f_k = k^n$ consists of the $(n+1)$ functions with values k^n, k^{n-1}, k^{n-2}, $\cdots$, k, 1.   See the discussion in F. B. Hildebrand, *Methods of Applied Mathematics*, Prentice-Hall, New York, 1952, pp. 242–245.   A proof that the method outlined is indeed successful in producing a solution may be found in T. Fort, *Finite Differences*, Oxford Univ. Press, London, 1948, pp. 127–128.   For a brief discussion of the method of undetermined coefficients as used to find solutions of differential equations, see H. W. Reddick and D. E. Kibbey, *Differential Equations*, 3rd ed., John Wiley, New York, 1956, pp. 148–156.

Table 3.1 indicates the form of the trial solution to be used corresponding to some simple functions r. If the right member r in (3.60) is the sum of several different functions, each function should be treated separately (see 5 of Problems 3.4). If the trial solution y^* includes a function which is a solution of the homogeneous equation, this y^* should first be multiplied by k and the new trial function used. If this function also contains a term which satisfies the reduced equation, then multiply again by k. For a second-order difference equation, one need go no further in seeking trial solutions. In each case, the constant coefficients must be determined so that y^* satisfies the complete difference equation for all k-values. One final example is given to illustrate this procedure.

TABLE 3.1

r_k in (3.60)	Trial Solution y_k^*
a^k	Aa^k
$\sin bk$ or $\cos bk$	$A \sin bk + B \cos bk$
k^n	$A_0 + A_1 k + A_2 k^2 + \cdots + A_n k^n$
$k^n a^k$	$a^k(A_0 + A_1 k + A_2 k^2 + \cdots + A_n k^n)$
$a^k \sin bk$ or $a^k \cos bk$	$a^k(A \sin bk + B \cos bk)$

Example

Find the general solution of

$$y_{k+2} - 4y_{k+1} + 4y_k = 3k + 2^k.$$

The auxiliary equation $m^2 - 4m + 4 = 0$ has the equal roots $m_1 = m_2 = 2$. The general solution of the reduced equation is therefore given by

$$Y_k = (C_1 + C_2 k)2^k.$$

To this we must add a particular solution of the complete equation. The right-hand term $3k$ leads to the trial solution $y_k^* = A_0 + A_1 k$; the term 2^k leads to the trial solution $A2^k$. But 2^k is a solution of the reduced equation. Therefore we multiply by k and try $Ak2^k$, but since this too is a solution of the reduced equation we must again multiply by k and so arrive at the trial solution $Ak^2 2^k$. Combining terms, we attempt a trial solution of the form

$$y_k^* = A_0 + A_1 k + Ak^2 2^k.$$

To determine the coefficients A_0, A_1, and A we perform the necessary calculation to find

$$y_{k+2}^* - 4y_{k+1}^* + 4y_k^* = (A_0 - 2A_1) + A_1 k + 8A2^k.$$

Thus if y^* is to satisfy the difference equation, we must have

$$A_0 - 2A_1 = 0 \qquad A_1 = 3 \qquad 8A = 1$$

or $A_0 = 6$, $A_1 = 3$, and $A = \frac{1}{8}$.

The general solution of the complete equation is therefore given by

$$y_k = (C_1 + C_2 k)2^k + 6 + 3k + \tfrac{1}{8}k^2 2^k.$$

Other methods, such as those involving an operator or a variation of parameters technique, are known for obtaining particular solutions of the complete equation.[4] Although of considerable theoretical interest, these techniques are fairly difficult to apply except for the very simplest right-hand functions r.   In any case, the method of undetermined coefficients suffices for our purposes.

PROBLEMS 3.4

1. Find particular solutions of the following difference equations by the method of undetermined coefficients:

(a) $y_{k+2} - 5y_{k+1} + 6y_k = 2$
(b) $8y_{k+2} - 6y_{k+1} + y_k = 2^k$
(c) $y_{k+2} - 3y_{k+1} + 2y_k = 1$
(d) $y_{k+2} - y_{k+1} - 2y_k = k^2$
(e) $y_{k+2} + y_k = \sin \dfrac{k\pi}{2}$
(f) $4y_{k+2} - 4y_{k+1} + y_k = 2$
(g) $y_{k+2} - 2y_{k+1} + y_k = 5 + 3k$
(h) $y_{k+2} - 2y_{k+1} + y_k = 2^k(k - 1)$

2. Find the general solutions of the difference equations in Problem 1.   (First find the general solutions of the corresponding homogeneous equations and then use Theorem 3.6.)

3. Find the solutions of the difference equations in Problem 1 which satisfy the initial conditions $y_0 = 1$ and $y_1 = -1$.   (*Hint:* Find appropriate values for the arbitrary constants in the general solutions found in Problem 2.)

4. For the difference equation in Problem 1(a), find the solution which satisfies the initial conditions $y_1 = -1$ and $y_2 = -9$.   Note that your answer is the solution obtained in Problem 3 for the initial conditions $y_0 = 1$ and $y_1 = -1$.   Why is this so?

5. Show that if $y^{(1)}$ is a solution of

$$y_{k+2} + a_1 y_{k+1} + a_2 y_k = r_k^{(1)}$$

and $y^{(2)}$ is a solution of

$$y_{k+2} + a_1 y_{k+1} + a_2 y_k = r_k^{(2)},$$

---

[4] A method of combining solutions of the homogeneous equation to obtain a particular solution of the complete equation is discussed (and applied) by T. Haavelmo, "The Inadequacy of Testing Dynamic Theory by Comparing Theoretical Solutions and Observed Cycles," *Econometrica*, 8 (1940), 312–321.   For the method of operators, see C. H. Richardson, *An Introduction to The Calculus of Finite Differences*, D. Van Nostrand, New York, 1954, pp. 111–117.

then $y^{(1)} + y^{(2)}$ is a solution of

$$y_{k+2} + a_1 y_{k+1} + a_2 y_k = r_k^{(1)} + r_k^{(2)}.$$

3.5 LIMITING BEHAVIOR OF SOLUTIONS

We now know how to go about obtaining the general solution of the complete second-order difference equation (3.60). But solving an equation is not enough. From the point of view of applications, in the social sciences and elsewhere, it is crucial to study the solution in order to obtain information about the function or functions which satisfy the difference equation.

When two initial values of y are prescribed, say y_0 and y_1, we have seen that the arbitrary constants in the general solution can be determined to find that unique solution of (3.60) which assumes the given values at $k = 0$ and $k = 1$. As in the case of the first-order difference equation considered in Section 2.6, this solution is a *sequence* $\{y_k\}$. In this section we study the variety of limiting behaviors exhibited by such solution sequences. The reader would do well to reread Sections 2.5 and 2.6 at this time. We shall use the notation and terminology introduced there.

The following examples illustrate the possible types of convergent and divergent sequences obtainable from second-order difference equations. Since our primary aim here is to study the solutions, we omit the details of their derivation. The reader may supply these by using the methods of the preceding sections.

Examples

(a) The homogeneous equation

$$y_{k+2} - 3y_{k+2} + 2y_k = 0$$

has general solution given by

$$Y_k = C_1 + C_2 2^k.$$

If the initial conditions $y_0 = 0$ and $y_1 = 1$ are prescribed, we obtain the solution

$$y_k = 2^k - 1.$$

The solution sequence $\{2^k - 1\}$ is $0, 1, 3, 7, 15, 31, \cdots$. It is clear that the sequence is of type D1 (diverges to $+\infty$).

If the initial conditions are changed to $y_0 = -1$ and $y_1 = -2$, Y_k specializes to the solution

$$y_k = -2^k.$$

Now the sequence $\{y_k\}$ is $-1, -2, -4, -8, -16, \cdots$, and thus diverges to $-\infty$ (type D2).

Finally, we obtain a sequence $\{y_k\}$ of type C1 if y_0 and y_1 are equal. For then the solution is the constant sequence $y_0, y_0, y_0, \cdots$.

(b) The difference equation

$$2y_{k+2} + 3y_{k+1} - 2y_k = 0$$

has general solution given by

$$Y_k = C_1(\tfrac{1}{2})^k + C_2(-2)^k.$$

If $y_0 = 1$ and $y_1 = \tfrac{1}{2}$, we find the particular solution y given by

$$y_k = (\tfrac{1}{2})^k.$$

The solution sequence $\{y_k\}$ is $1, \tfrac{1}{2}, \tfrac{1}{4}, \tfrac{1}{8}, \cdots$ and is therefore bounded and monotone decreasing with limit 0. It is of type C3.

If the initial conditions are $y_0 = 1$ and $y_1 = -2$, the solution becomes

$$y_k = (-2)^k,$$

or $1, -2, 4, -8, 16, -32, \cdots$. This solution sequence is divergent and oscillates infinitely (type D4).

Finally, we note that if the initial conditions are $y_0 = 2$ and $y_1 = -\tfrac{3}{2}$, we obtain the solution

$$y_k = (\tfrac{1}{2})^k + (-2)^k,$$

or $2, \tfrac{1}{2} - 2, \tfrac{1}{4} + 4, \tfrac{1}{8} - 8, \tfrac{1}{16} + 16, \cdots$. The sequence $\{y_k\}$ is the sum of the two sequences $\{(\tfrac{1}{2})^k\}$ and $\{(-2)^k\}$. The second of these dominates the first and the sum again oscillates infinitely.

(c) The difference equation

$$y_{k+2} + y_k = 0$$

has the general solution given by

$$Y_k = A \cos\left(\frac{k\pi}{2} + B\right).$$

With initial conditions $y_0 = 1$ and $y_1 = 0$, we get the particular solution

$$y_k = \cos\frac{k\pi}{2}.$$

But

$$\cos\frac{k\pi}{2} = \begin{cases} 1 & \text{if } k = 0, 4, 8, 12, \cdots \\ -1 & \text{if } k = 2, 6, 10, 14, \cdots \\ 0 & \text{if } k = 1, 3, 5, 7, \cdots \end{cases}$$

so the solution sequence $\{y_k\}$ is $1, 0, -1, 0, 1, 0, -1, 0, \cdots$. This sequence is divergent and is of type D3 (oscillates finitely). Note the periodicity of this sequence: each term is repeated four steps further along.

(d) The difference equation

$$4y_{k+2} + y_k = 0$$

has the general solution given by

$$Y_k = A(\tfrac{1}{2})^k \cos\left(\frac{k\pi}{2} + B\right).$$

If $y_0 = 1$ and $y_1 = 0$ are prescribed, we obtain the special solution

$$y_k = (\tfrac{1}{2})^k \cos\frac{k\pi}{2}$$

with terms $1, 0, -\tfrac{1}{4}, 0, \tfrac{1}{16}, 0, -\tfrac{1}{64}, 0, \tfrac{1}{256}, \cdots$. This sequence is again oscillatory, as in the preceding example, but now the factor $(\tfrac{1}{2})^k$ creates damped oscillations. The solution sequence is convergent and of type C4.

With these illustrative examples before us, a number of general remarks may be made: (1) the solution sequence, if convergent, may remain constant, steadily increase or decrease, or exhibit damped oscillations; if divergent, it may steadily increase (to $+\infty$), steadily decrease (to $-\infty$), or exhibit steady (finite) or explosive (infinite) oscillations; (2) the behavior of the solution sequence depends upon *both* the difference equation and the initial conditions; and (3) the roots of the auxiliary equation are important factors in the determination of the limiting behavior of the solution.

To study these matters in greater detail, we first concentrate on the homogeneous difference equation and consider separately the three forms for the general solution (as given in Theorem 3.7).

CASE 1.   REAL ROOTS, $m_1 \neq m_2$

Suppose that the root with the greater absolute value has been named m_1, i.e., $|m_1| > |m_2|$. Then *the limiting behavior of the solution sequence* $\{C_1 m_1{}^k + C_2 m_2{}^k\}$ *is the same as that of* $\{C_1 m_1{}^k\}$, *provided* $C_1 \neq 0$. To prove this, write

$$C_1 m_1{}^k + C_2 m_2{}^k = m_1{}^k \left[C_1 + C_2 \left(\frac{m_2}{m_1}\right)^k \right]$$

and note that the quantity in brackets approaches C_1 as $k \to \infty$ since $-1 < m_2/m_1 < 1$. (This statement follows formally from Theorems 2.5 and 2.6.) Then

$$\frac{C_1 m_1{}^k}{C_1 m_1{}^k + C_2 m_2{}^k} = \frac{C_1}{C_1 + C_2 \left(\frac{m_2}{m_1}\right)^k} \to \frac{C_1}{C_1} = 1 \quad \text{as } k \to \infty$$

so that the ratio of $C_1 m_1{}^k$ to $C_1 m_1{}^k + C_2 m_2{}^k$ is arbitrarily close to 1 for sufficiently large k-values. Their limiting behavior is then seen to be identical.

Thus, if $C_1 \neq 0$, it suffices to study the sequence $\{C_1 m_1{}^k\}$. But this has already been done in detail in Theorems 2.5 and 2.6. We mention only some of the more important results: (1) if $|m_1| \leq 1$, then the sequence converges; (2) if $|m_1| > 1$, then it diverges; (3) if $-1 < m_1 < 0$, it is damped oscillatory; and (4) if $m_1 < -1$, it oscillates infinitely.

If $C_1 = 0$, then the solution becomes $\{C_2 m_2{}^k\}$ and similar considerations must be applied to this sequence. The condition $C_1 = 0$ means, for the homogeneous equation, that the solution has the value C_2 at $k = 0$ and $C_2 m_2$ at $k = 1$. This exceptional case arises whenever the initial values are so related. [Cf. Example (a) above.]

CASE 2.   REAL ROOTS, $m_1 = m_2$

The solution sequence is now $\{(C_1 + C_2k)m_1{}^k\}$. This certainly diverges if $|m_1| > 1$ (unless both C_1 and C_2 are 0), and also diverges if $|m_1| = 1$ (unless $C_2 = 0$). Now consider the case $|m_1| < 1$. Then it can be proved[5] that the sequence $\{km_1{}^k\}$ converges to 0. Hence, if $|m_1| < 1$, the solution sequence also converges to 0. As in the preceding case, oscillatory behavior is exhibited if m_1 is negative.

CASE 3.   COMPLEX ROOTS

The solution sequence is $\{Ar^k \cos (k\theta + B)\}$ and we assume $A \neq 0$. The reader should recall from trigonometry that the function given by $\cos (k\theta + B)$ is a periodic function, with values between -1 and $+1$. The *period* is $(2\pi/\theta)$, the *amplitude* is 1. The sequence $\{\cos (k\theta + B)\}$ oscillates finitely. This steady oscillation is unchanged, damped, or magnified, in the solution sequence, as the factor r^k is equal to, less than, or greater than 1. As examples, consider the sequences

$$\left\{\cos \frac{k\pi}{2}\right\} = 1, 0, -1, 0, 1, 0, -1, 0, \cdots$$

$$\left\{(\tfrac{1}{2})^k \cos \frac{k\pi}{2}\right\} = 1, 0, -\tfrac{1}{4}, 0, \tfrac{1}{16}, 0, -\tfrac{1}{64}, 0, \cdots$$

$$\left\{2^k \cos \frac{k\pi}{2}\right\} = 1, 0, -4, 0, 16, 0, -64, 0, \cdots$$

In general, it can be shown that the solution sequence converges to 0 if $0 < r < 1$ and diverges if $r \geq 1$. But the sequence is always oscillatory due to the presence of the cosine function.

The foregoing analysis is basic to the complete characterization of the solution sequences of the homogeneous equation. A classification[6] of types of solutions is cumbersome owing to the exceptional initial values for which one or both of the arbitrary constants C_1 and C_2 are 0. Additional difficulties are introduced when one studies the complete difference equation, for the right-hand member may create a divergent solution when the solution of the corresponding homogeneous equation is convergent. For example, the solution of the homogeneous equation

$$8y_{k+2} - 6y_{k+1} + y_k = 0$$

[5] Actually $k^n m_1{}^k \to 0$ for any positive integer n, if $|m_1| < 1$. Although $k^n \to \infty$, the factor $m_1{}^k \to 0$ and this latter effect predominates. See G. H. Hardy, *A Course of Pure Mathematics*, 9th ed., Macmillan, New York, 1949, p. 141.

[6] See W. J. Baumol, *Economic Dynamics*, Macmillan, New York, 1951, pp. 196–204.

is given by

$$Y_k = C_1(\tfrac{1}{2})^k + C_2(\tfrac{1}{4})^k$$

and is therefore convergent. But the complete equation

$$8y_{k+2} - 6y_{k+1} + y_k = 2^k$$

has the general solution given by

$$y_k = Y_k + \frac{2^k}{21}$$

and is therefore divergent.

We shall return to nonhomogeneous equations and their solution sequences in Section 4.1 where we take up the *stability* of such equations.

In spite of the multitude of possible behaviors for the homogeneous equation, in one case we may be certain, *whatever the initial values*, the solution sequence will converge to 0. We shall need the following theorem in Section 4.1. Its proof is implicitly contained in the results already stated for the homogeneous equation.

We first define the number

$$\rho = \max\left(|m_1|, |m_2|\right)$$

which is the larger of $|m_1|$ and $|m_2|$ if they are different, and either $|m_1|$ or $|m_2|$ if they are equal. ("max" stands for "maximum".) Thus if $m_1 = \tfrac{1}{2}$ and $m_2 = 3$, then $\rho = 3$; if $m_1 = -2$ and $m_2 = 1$, then $\rho = 2$; if $m_1 = m_2 = -3$, then $\rho = 3$. If one defines the absolute value of a complex number as its modulus when written in polar form, then when m_1 and m_2 are the complex conjugate roots given by $r(\cos\theta \pm i\sin\theta)$, we have $|m_1| = |m_2| = r$ and therefore $\rho = r$. Since no root of the auxiliary equation of the second-order equation is zero, we know that $\rho > 0$.

THEOREM 3.8. *Let*

$$\rho = \max\left(|m_1|, |m_2|\right),$$

where m_1 and m_2 are the roots of the auxiliary equation of the homogeneous second-order difference equation

$$y_{k+2} + a_1 y_{k+1} + a_2 y_k = 0.$$

Then $\rho < 1$ is a necessary and sufficient condition for the solution sequence $\{y_k\}$ to converge with limit 0 for all initial values y_0 and y_1.

PROBLEMS 3.5

1. In 2 of Problems 3.3, particular solutions of six difference equations were obtained. (i) Determine the limiting behavior of each solution. (ii) Calculate

ρ for each difference equation and show that the condition $\rho < 1$ is satisfied for only one of these equations.

2. Consider the nonhomogeneous difference equations in 1 of Problems 3.4. The general solutions of the corresponding homogeneous equations were found in 2 of Problems 3.4. (i) Find the particular solutions of these homogeneous equations which satisfy the initial conditions $y_0 = 1$ and $y_1 = -1$. (ii) Determine the limiting behavior of each solution.

In 3 of Problems 3.4, solutions of the complete equations were found subject to these same initial conditions. (iii) Determine the limiting behavior of each solution of the complete equation and compare with the behavior of the solution of the corresponding homogeneous equation.

3. Consider the equation

$$y_{k+2} - \alpha y_{k+1} + \alpha y_k = 0,$$

where α is a positive constant. Solve the equation for $\alpha = \frac{1}{2}$, $\alpha = 1$, and $\alpha = 2$ and show that the solutions are oscillatory, but with damped, finite, and infinite oscillations respectively.

3.6 ILLUSTRATIVE EXAMPLES FROM THE SOCIAL SCIENCES

Example 1. National Income

In Chapter 0, following Samuelson (see the reference cited), we derived the difference equation[7] (0.10) for the national income function Y. With Y_t denoting the national income in period t, and α and β positive constants (α being the marginal propensity to consume and β the relation), the difference equation is

$$Y_t = \alpha(1 + \beta)Y_{t-1} - \alpha\beta Y_{t-2} + 1 \qquad t = 2, 3, 4, \cdots.$$

We rewrite this equation in the equivalent standard form

$$(3.65) \qquad Y_{t+2} - \alpha(1 + \beta)Y_{t+1} + \alpha\beta Y_t = 1 \qquad t = 0, 1, 2, \cdots.$$

The national income Y is determined as a solution of this second-order nonhomogeneous difference equation with constant coefficients. As soon as two initial values, Y_0 and Y_1, are specified, the income is determined for all later periods by (3.65).

[7] For a discussion of the economic significance of this and related mathematical models, see W. Fellner, *Trends and Cycles in Economic Activity*, Henry Holt, New York, 1956, pp. 308–338. A large variety of mathematical models in economics, including this interaction model, are discussed in E. F. Beach, *Economic Models*, John Wiley, New York, 1957.

In order to compare the present analysis with the arithmetical one summarized in Table 0.1, we treat the same two special cases considered in Chapter 0. The initial values are

$$(3.66) \qquad Y_0 = 2 \quad \text{and} \quad Y_1 = 3.$$

(Note that we are starting with $t = 0$, but in Chapter 0 it was more natural to start with $t = 1$. Thus the present Y_0 and Y_1 are the Y_1 and Y_2 of Table 0.1 and all other values must be similarly relabeled.)

Now consider the difference equation (3.65) in the special case when $\alpha = 0.5$ and $\beta = 1$. It reduces to

$$(3.67) \qquad Y_{t+2} - Y_{t+1} + \tfrac{1}{2} Y_t = 1.$$

We solve this equation by first finding the general solution of the corresponding homogeneous equation and then adding to that a particular solution of the complete equation. The homogeneous equation has

$$m^2 - m + \tfrac{1}{2} = 0$$

as its auxiliary equation. By the quadratic formula, we find the roots

$$m_1 = \tfrac{1}{2} + \tfrac{1}{2}i \qquad m_2 = \tfrac{1}{2} - \tfrac{1}{2}i.$$

We use (3.45) and (3.46) to transform these complex numbers to polar form:

$$r = \sqrt{(\tfrac{1}{2})^2 + (\tfrac{1}{2})^2} = 1/\sqrt{2} \qquad \cos \theta = \sqrt{2}/2 \qquad \sin \theta = \sqrt{2}/2.$$

Hence, $\theta = \pi/4$ and the polar form of the roots is

$$\frac{1}{\sqrt{2}} \left(\cos \frac{\pi}{4} \pm \sin \frac{\pi}{4} \right).$$

By Theorem 3.7 the solution of the homogeneous equation is given by (3.56) in the form

$$(3.68) \qquad Y_t = A \left(\frac{1}{\sqrt{2}} \right)^t \cos \left(\frac{t\pi}{4} + B \right).$$

The problem of finding a particular solution of (3.67) is simple. We assume a trial solution of the form $y_t^* = k$, a constant. If this is to satisfy (3.67) we must take $k = 2$. The general solution of (3.67) is therefore given by

$$(3.69) \qquad Y_t = A \left(\frac{1}{\sqrt{2}} \right)^t \cos \left(\frac{t\pi}{4} + B \right) + 2.$$

The initial conditions (3.66) enable us to determine the constants A and B. Putting $t = 0$ and $t = 1$ in (3.69) we find

$$A \cos B = 0 \qquad A \frac{1}{\sqrt{2}} \cos \left(\frac{\pi}{4} + B \right) = 1.$$

The first of these is satisfied if we take $B = \pi/2$. Since $\cos (3\pi/4) = -(1/\sqrt{2})$, the second equation requires that $A = -2$. Thus, the particular solution of (3.67) which satisfies the initial conditions (3.66) is given by

$$Y_t = -2 \left(\frac{1}{\sqrt{2}} \right)^t \cos \left(\frac{t\pi}{4} + \frac{\pi}{2} \right) + 2.$$

This may be simplified by noting that $\cos [x + (\pi/2)] = -\sin x$. Therefore

(3.70)
$$Y_t = 2 \left(\frac{1}{\sqrt{2}} \right)^t \sin \frac{t\pi}{4} + 2.$$

The presence of the sine term makes Y an oscillating function of time. Since $r = 1/\sqrt{2} < 1$, the oscillations are damped and the first term on the right in (3.70) approaches zero as t increases. The constant term, 2, remains and the sequence $\{ Y_t \}$ converges to this limit, undergoing damped oscillations above and below the level $Y_t = 2$. Thus, "A constant continuing level of government expenditure will result in damped oscillatory movements of national income, gradually approaching the asymptote $1/(1 - \alpha)$ times the constant level of government expenditure." (Samuelson, p. 267.) We have just considered the special case of this result when $\alpha = 0.5$. [The constant level of government expenditure was assumed in (0.9) to be 1.]

The case $\alpha = 0.8$ and $\beta = 2$ is treated in much the same way. Now (3.65) becomes

(3.71)
$$Y_{t+2} - 2.4 Y_{t+1} + 1.6 Y_t = 1.$$

The auxiliary equation

$$m^2 - 2.4m + 1.6 = 0$$

has the two roots

$$m_1 = 1.2 + 0.4i \qquad \text{and} \qquad m_2 = 1.2 - 0.4i,$$

as may be calculated from the quadratic formula. Since the roots are complex, we are certain of oscillatory movement of national income. To find out if these oscillations are damped, steady, or explosive, we calculate the modulus r:

$$r = \sqrt{(1.2)^2 + (0.4)^2} = \sqrt{1.6}.$$

Since $r > 1$, the oscillations are unbounded. The general solution of (3.71) can be obtained, but it is unnecessary for the conclusion that in this case we have "explosive, ever-increasing oscillations around an asymptote computed as above."

This difference equation (3.65) will be studied again in Section 4.1 with reference to the stability of its equilibrium value and the dependence of income behavior on the values of the parameters α and β. (Cf. 1 and 2 of Problems 3.6.)

Example 2. The Transmission of Information[8]

Imagine a signaling system that has only two signals, S_1 and S_2 (e.g., the dots and dashes in telegraphy). Messages are transmitted over some channel by first coding them into sequences of these two signals. Suppose S_1 requires exactly t_1 units of time and S_2 exactly t_2 units of time to be transmitted. Let N_t denote the number of possible message sequences of duration t.

Consider first those messages of duration t which end with an S_1. Since S_1 takes t_1 units of time, this last signal must start at time $t - t_1$. Up to time $t - t_1$ there are N_{t-t_1} possible messages to which the last S_1 may be appended. Hence, the total number of messages of duration t which end with an S_1 is just N_{t-t_1}. Similarly, the total number of messages of duration t which end with an S_2 is given by N_{t-t_2}. But a message of duration t must end in either an S_1 or an S_2. Hence

$$(3.72) \qquad N_t = N_{t-t_1} + N_{t-t_2}.$$

We consider the special case $t_1 = 1$ and $t_2 = 2$, i.e., where one signal takes twice as long to be transmitted over the channel as the other. Then

$$(3.73) \qquad N_t = N_{t-1} + N_{t-2}$$

which we can rewrite in the standard form

$$(3.74) \qquad N_{t+2} - N_{t+1} - N_t = 0$$

with t-values $0, 1, 2, \cdots$

This is a linear homogeneous second-order difference equation with constant coefficients. Its auxiliary equation is $m^2 - m - 1 = 0$ which

[8] C. E. Shannon and W. Weaver, *The Mathematical Theory of Communication*, Univ. of Illinois Press, Urbana, 1949, pp. 7–8. A bibliography of papers on information theory of direct interest to psychologists may be found in G. A. Miller, "What is Information Measurement?" *American Psychologist*, 8 (1953), 3–11. A more up-to-date and complete bibliography is in R. D. Luce, *A Survey of the Theory of Selective Information and Some of its Behavioral Applications*, Columbia Univ. Bureau of Applied Social Research, New York, June 1956 (hectographed).

has the roots $(1 \pm \sqrt{5})/2$. These are real, unequal roots so, by Theorem 3.7, the general solution of (3.74) is given by

$$(3.75) \qquad N_t = C_1 \left(\frac{1 + \sqrt{5}}{2} \right)^t + C_2 \left(\frac{1 - \sqrt{5}}{2} \right)^t.$$

If initial values N_0 and N_1 are specified, constants C_1 and C_2 can be determined and the required solution found. In view of the meaning of N_t, we take $N_0 = 0$ and $N_1 = 1$ and obtain the particular solution with the constants in (3.75) given by (cf. 3 of Problems 3.6)

$$C_1 = 1/\sqrt{5} \qquad C_2 = -1/\sqrt{5}.$$

The *capacity* C of the channel is defined by Shannon as follows:

$$(3.76) \qquad C = \lim_{t \to \infty} \frac{\log_2 N_t}{t},$$

where $\log_2$ denotes the logarithm to the base 2. (The base 2, rather than the more common bases 10 or $e = 2.718 \cdots$, is used in information theory. This is due to the following definition: if a single selection is to be made from a number of equally probable alternatives, and if information is transmitted which reduces the number of alternatives by a factor of 2, this amount of information is 1 *bit*, the unit of information. For example, 1 bit is transmitted in reducing the number of alternatives from 32 to 16, from 16 to 8, from 8 to 4, etc. Since $\log_2 2^n = n$, we have $\log_2 32 = 5$, $\log_2 16 = 4$, $\log_2 8 = 3$, $\log_2 4 = 2$. The logarithm to the base 2 of the number of alternatives varies in steps equal to the amount of information transmitted.)

To calculate the capacity C, we first note, according to the argument in Section 3.5, that the sequence $\{N_t\}$ has the same limiting behavior as $\{C_1[(1 + \sqrt{5})/2]^t\}$. This follows from the observation that the root $m_1 = (1 + \sqrt{5})/2$ ($= 1.6$, approximately) is greater in absolute value than the root $m_2 = (1 - \sqrt{5})/2$ ($= -0.6$, approximately) so the first term on the right in (3.75) predominates, the second approaching zero as t increases. Hence, from (3.76),

$$C = \lim_{t \to \infty} \frac{\log_2 C_1 m_1^{\,t}}{t}.$$

But

$$\log_2 C_1 m_1^{\,t} = \log_2 C_1 + t \log_2 m_1$$

so that

$$C = \lim_{t \to \infty} \left(\frac{\log_2 C_1}{t} + \log_2 m_1 \right) = \log_2 m_1.$$

We find C is approximately 0.7.

General results relating to the difference equation (3.72) and to the capacities of channels which allow the transmission of more complex message sequences may be found in the Shannon work cited in footnote 8.

Example 3. Inventory Analysis

It would be well to reread Section 2.9 at this time for we now are able to complete the discussion initiated there of Metzler's "pure inventory cycle."

The income function y was shown, with assumptions given in the previous discussion, to satisfy the second-order difference equation (2.88) which we repeat here:

$$(3.77) \qquad y_{t+2} - 2\beta y_{t+1} + \beta y_t = v_0.$$

The constant β (the marginal propensity to consume) is restricted by the condition

$$(3.78) \qquad 0 < \beta < 1$$

and v_0 is the constant level of noninduced investment.

We now find the solution of (3.77) and show that the sequence $\{y_t\}$ of income values will undergo damped oscillatory behavior around a limit determined by the values of β and v_0.

The auxiliary equation of the homogeneous difference equation corresponding to (3.77) is

$$m^2 - 2\beta m + \beta = 0$$

with roots (obtained using the quadratic formula)

$$(3.79) \qquad \beta \pm \sqrt{\beta^2 - \beta}.$$

Since $0 < \beta < 1$, the quantity $\beta^2 - \beta$ under the square root sign is negative and the roots are the complex numbers

$$(3.80) \qquad \beta \pm i\sqrt{\beta(1 - \beta)}.$$

We use (3.45) and (3.46) to write these complex conjugate roots in polar form. Their modulus r is given by

$$(3.81) \qquad r = \sqrt{\beta^2 + \beta(1 - \beta)} = \sqrt{\beta}$$

and θ is that angle between $-\pi$ and π for which

$$\cos \theta = \frac{\beta}{r} = \sqrt{\beta} \qquad \sin \theta = \frac{\sqrt{\beta(1 - \beta)}}{r} = \sqrt{1 - \beta}.$$

Then the solution of the homogeneous equation is, by Theorem 3.7,

$$(3.82) \qquad Y_t = A(\sqrt{\beta})^t \cos{(t\theta + B)},$$

where A and B are arbitrary constants.

To find a particular solution of (3.77) we use a trial solution of the form $y_t{}^* = k$, a constant. If this is to satisfy (3.77), we must have

$$k - 2\beta k + \beta k = v_0$$

so a particular solution is given by

$$(3.83) \qquad y_t{}^* = \frac{v_0}{1 - \beta}.$$

Thus, the general solution of the national income equation (3.77) has the form

$$(3.84) \qquad y_t = A(\sqrt{\beta})^t \cos{(t\theta + B)} + \frac{v_0}{1 - \beta}.$$

The cosine term produces cyclical fluctuations because of its oscillation between positive and negative values. These fluctuations are damped by the factor $(\sqrt{\beta})^t$ since $\sqrt{\beta} < 1$. Thus, the first term on the right in (3.84) converges to 0 and y_t fluctuates about and approaches the constant equilibrium value $v_0/(1 - \beta)$ (cf. 6 of Problems 3.6).

It is possible to derive a simple equation showing the relation of inventory levels to national income. We proceed as follows, using the notation introduced in Section 2.9. The inventory level (i_t) at the end of period t is equal to the inventory level (i_{t-1}) one period before plus the amount (s_t) produced for inventories in period t, plus the excess of consumers' goods produced for sale (u_t) over goods actually sold (βy_t) in period t. That is,

$$i_t = i_{t-1} + s_t + u_t - \beta y_t.$$

But by (2.84),

$$s_t + u_t = y_t - v_0$$

so that

$$i_t = i_{t-1} + (1 - \beta)y_t - v_0.$$

We write this in the more useful form

$$(3.85) \qquad \Delta i_{t-1} = i_t - i_{t-1} = (1 - \beta)\left(y_t - \frac{v_0}{1 - \beta}\right).$$

The left-hand side of (3.85) is the amount by which the inventory level changes during period t. The right-hand side, except for the positive constant $(1 - \beta)$, is the difference between the actual income in period t

and the limiting equilibrium level of income.   We know from the solution
(3.84) that y_t oscillates around this limiting value.   But (3.85) shows that,
whenever income exceeds $v_0/(1 - \beta)$, whether it is *falling* toward this value
or *rising* away from it, the change in inventory level will be positive.   Thus
inventories lag behind the movement of income, additions to inventory
being made even after income has reached its peak.   Similarly, whenever
income is below $v_0/(1 - \beta)$, whether falling further below or rising toward
this value, the inventory level decreases.   Inventories again lag, sub-
tractions from inventory stock being made even after income has turned
upward from its lowest point[9] (cf. 7 of Problems 3.6).

'In this "pure inventory cycle" model, expectation of future sales depends
only on the *level* of past sales.   It is more reasonable to think of expectations as also depending on the *direction* in which sales have moved.   For
future expectations may be different, depending upon whether a certain
level of sales has been reached by a *decrease* from a higher or an *increase*
from a lower level.   To take account of this fact, Metzler introduces a
coefficient of expectation, η, which measures the extent to which an
observed change in level of sales in any period is expected to continue in
the next period (cf. 9 of Problems 3.6).   We shall not pursue this any
further here, except to note that the difference equation which Metzler
derives and studies becomes

$$(3.86) \qquad y_{t+2} - \beta(1 + \eta)y_{t+1} + \eta\beta y_t = v_0.$$

[Note that this reduces to (3.77) when $\eta = 1$, this value of the coefficient
of expectation implying that a given change in sales is expected to continue
unaltered.   In general, the parameter η is between -1 and 1.]

A comparison of (3.86) with the Samuelson equation (3.65) shows that
the corresponding homogeneous equations are identical except that the
parameters α and β in (3.65) are replaced by β and η respectively.
Thus, our study of the Samuelson equation in Section 4.1 can be directly
applied to this inventory cycle model.

PROBLEMS 3.6

1. Consider (3.65) in the case where $\alpha = 0.5$ and $\beta = 0$.   Show that the
equation reduces to a nonhomogeneous first-order difference equation.   Find
the solution for which $Y_1 = 3$ and discuss the behavior of the sequence $\{Y_t\}$.

---

[9] For a discussion of the so-called Lundberg and Robertson lags in dynamic economics,
see L. A. Metzler, "Three Lags in the Circular Flow of Income," in *Income, Employment
and Public Policy*, essays in honor of Alvin H. Hansen, W. W. Norton, New York,
1948, pp. 11–32.

2. Consider (3.65) with $\alpha = 0.8$ and $\beta = 4$. Show that one obtains an "ever-increasing national income, eventually approaching a compound interest rate of growth," except if $Y_0 = Y_1 = 5$, when $Y_t = 5$ for all t.

3. A sequence of positive integers is determined by requiring that each term (after the first two) of the sequence is the sum of the two preceding terms. Let y_k denote the kth term ($k = 0, 1, 2, \cdots$). The values $y_0 = 0$ and $y_1 = 1$ are prescribed. (This sequence, with terms $0, 1, 2, 3, 5, 8, 13, \cdots$, is the so-called *Fibonacci sequence*.) Show that $y_{k+2} = y_{k+1} + y_k$ so that y satisfies the same difference equation as the function N considered in text Example 2. Find the solution subject to the given initial conditions and thus show that term number k in the Fibonacci sequence is

$$y_k = \frac{1}{\sqrt{5}}\left[\left(\frac{1+\sqrt{5}}{2}\right)^k - \left(\frac{1-\sqrt{5}}{2}\right)^k\right].$$

4. Consider (3.72) in the special case $t_1 = t_2 = 1$, i.e., where both signals S_1 and S_2 take 1 unit of time for transmission over the channel. Show that $N_{t+1} = 2N_t$ ($t = 1, 2, 3, \cdots$) and solve with the initial condition $N_1 = 2$. Show that the capacity of the channel is $C = 1$.

5. Consider (3.72) in the special case $t_1 = t_2 = 2$, i.e., where both signals S_1 and S_2 take 2 units of time for transmission over the channel. Show that $N_{t+2} = 2N_t$ ($t = 2, 3, 4, \cdots$) and solve with the initial conditions $N_2 = 2$, $N_3 = 2$. Show that the channel capacity is $C = \frac{1}{2}$.

6. Follow the analysis in the text, starting with (3.77), assuming $\beta = \frac{1}{2}$ and $v_0 = 500$. If the initial values $y_0 = 1000$ and $y_1 = 1100$ are prescribed, show that the national income in period t is given by

$$y_t = 200\left(\frac{1}{\sqrt{2}}\right)^t \sin\frac{t\pi}{4} + 1000.$$

Calculate the values $y_2, y_3, \cdots, y_8$ and note the damped oscillations.

7. Using the data of Problem 6, show that (3.85) reduces to

$$i_t - i_{t-1} = \tfrac{1}{2}(y_t - 1000).$$

Assume an initial inventory level $i_0 = 500$ and use this equation and the values $y_1, \cdots, y_8$ obtained in Problem 6 to calculate the inventory levels $i_1, \cdots, i_8$. Note the way in which inventory lags behind income.

8. In Example 3, show that if β, the marginal propensity to consume, is greater than 1, the income function y of (3.77) does not oscillate. If initial values are excluded for which the arbitrary constants in the solution of the homogeneous equation become 0, show that the sequence $\{y_t\}$ diverges to $\pm\infty$ in this case. [*Hint:* Both roots (3.79) of the auxiliary equation are positive and the larger is greater than 1.]

9. Suppose levels of sales in periods $t - 1$ and $t - 2$ are known and sales in period t are to be estimated. Metzler defines the coefficient of expectation, η, as follows:

$$\eta = \frac{\text{expected change of sales between periods } t \text{ and } t - 1}{\text{observed change of sales between periods } t - 1 \text{ and } t - 2}.$$

Show (a) if $\eta = 0$, then a given level of sales in period $t - 1$ is expected to be maintained in period t; (b) if $\eta = 1$, sales in period t are expected to differ from sales in period $t - 1$ in precisely the same way as sales in period $t - 1$ differed from sales in period $t - 2$; (c) if $\eta = -1$, an observed change of sales between periods $t - 1$ and $t - 2$ is expected to be temporary so that sales in period t are expected to return to their level in period $t - 2$.

If sales equal 120 in period $t - 1$ and sales equal 100 in period $t - 2$, calculate anticipated sales in period t if $\eta = 0, \frac{1}{2}, 1, -\frac{1}{2}$, and -1.

10. In 13 of Problems 2.7, the difference equation

$$A_{k+1} = (1 + i)A_k + (k + 1)R,$$

with $A_0 = 0$, was derived for the amount of an increasing annuity. Obtain the solution given there by using Theorem 3.3. (Find a particular solution of the complete equation by the method of undetermined coefficients.)

11. Suppose

$$p_{t+2} = Kr_0 + K\alpha(p_{t+1} - p_t),$$

where K, α, and r_0 are positive constants. Find the general solution and determine its limiting behavior if (a) $K\alpha = 1$; (b) $K\alpha = 2$, (c) $K\alpha = \frac{1}{2}$. (See 8 of Problems 2.8 and footnote 15 cited there.)

12. Consider the difference equation

$$Y_{t+2} - \frac{C}{s} Y_{t+1} + \frac{C}{s} Y_t = 0,$$

where C and s are positive constants and Y_t is warranted income in period t. By studying the roots of the auxiliary equation, show[10] that "For $C \geq 4s$ this gives an explosive time path for warranted income $\cdots$, but for $C < 4s$ warranted income behaves cyclically." (Cf. 6 of Problems 2.8.)

13. Consider the difference equation[11]

$$P_{n+2} - (2 - A)P_{n+1} + (1 - A + B)P_n = 0,$$

where A and B are constants. If $P_0 = 0$ and $P_1 = p_e$ are prescribed, show that

$$P_n = \frac{p_e}{\alpha} (r_1{}^n - r_2{}^n),$$

where

$$\alpha = \sqrt{A^2 - 4B}, \qquad r_1 = \frac{2 - A + \alpha}{2}, \qquad r_2 = \frac{2 - A - \alpha}{2}.$$

---

[10] W. J. Baumol, "Formalisation of Mr. Harrod's Model," *Economic Journal, 59* (1949), 625–629.

[11] R. R. Bush and J. W. M. Whiting, "On the Theory of Psychoanalytic Displacement," *Journal of Abnormal and Social Psychology, 48* (1953), 261–272. We have written $A = p + p_e + p_w$ and $B = pp_e$ in their equation [15], p. 268.

3.7 THE GENERAL CASE OF ORDER n

The theory of the linear nth-order difference equation with constant coefficients

$$(3.87) \qquad y_{k+n} + a_1 y_{k+n-1} + \cdots + a_{n-1} y_{k+1} + a_n y_k = r_k$$

is a straightforward extension of that already developed in the special case $n = 2$. We outline the general results without giving proofs. In each case, the proof assumes precisely the same form as the corresponding proof for the second-order equation.

The homogeneous equation

$$(3.88) \qquad y_{k+n} + a_1 y_{k+n-1} + \cdots + a_{n-1} y_{k+1} + a_n y_k = 0$$

is studied first. The *auxiliary equation* is now

$$(3.89) \qquad m^n + a_1 m^{n-1} + \cdots + a_{n-1} m + a_n = 0,$$

which is an algebraic equation of degree n. It has exactly n roots, denoted by $m_1, m_2, \cdots, m_n$. These may be real or complex numbers, and any root may be distinct from the others or may be a repeated root.

A *fundamental set* of solutions is now defined as a set of n solutions $y^{(1)}, y^{(2)}, \cdots, y^{(n)}$, for which the nth order determinant[12]

$$\begin{vmatrix} y_0^{(1)} & y_0^{(2)} & \cdots & y_0^{(n)} \\ y_1^{(1)} & y_1^{(2)} & \cdots & y_1^{(n)} \\ \cdot & \cdot & \cdots & \cdot \\ y_{n-1}^{(1)} & y_{n-1}^{(2)} & \cdots & y_{n-1}^{(n)} \end{vmatrix}$$

is different from zero. [When $n = 2$, this reduces to the 2×2 determinant (3.26).] Such a fundamental set of solutions is obtained as follows (the A's, B's, and C's are arbitrary constants):

(1) For each real unrepeated root m, write the solution (containing one arbitrary constant)

$$C_1 m^k.$$

---

[12] A definition of a determinant of order n may be found in L. E. Dickson, *New First Course in the Theory of Equations*, John Wiley, New York, 1939, pp. 110–111. From this definition it is shown that a determinant of order n may be evaluated in terms of determinants of order $(n - 1)$. When $n = 3$, this yields the equation

$$\begin{vmatrix} y_0^{(1)} & y_0^{(2)} & y_0^{(3)} \\ y_1^{(1)} & y_1^{(2)} & y_1^{(3)} \\ y_2^{(1)} & y_2^{(2)} & y_2^{(3)} \end{vmatrix} = y_2^{(1)} \begin{vmatrix} y_0^{(2)} & y_0^{(3)} \\ y_1^{(2)} & y_1^{(3)} \end{vmatrix} - y_2^{(2)} \begin{vmatrix} y_0^{(1)} & y_0^{(3)} \\ y_1^{(1)} & y_1^{(3)} \end{vmatrix} + y_2^{(3)} \begin{vmatrix} y_0^{(1)} & y_0^{(2)} \\ y_1^{(1)} & y_1^{(2)} \end{vmatrix}$$

The reader can use this equation, together with the definition of second-order determinants in (3.26), to check that the solutions obtained for the difference equations of order 3 in Example (*a*) of this section and 1 of Problems 3.7 do indeed form fundamental sets.

(2) If a real root m is repeated p times, write the solution (containing p arbitrary constants)

$$(C_1 + C_2k + C_3k^2 + \cdots + C_pk^{p-1})m^k.$$

(3) For each pair of unrepeated complex conjugate roots with modulus r and amplitude θ, write the solution (containing two arbitrary constants)

$$Ar^k \cos(k\theta + B).$$

(4) If a pair of complex conjugate roots is repeated p times, write the solution (containing $2p$ arbitrary constants)

$$r^k[A_1 \cos(k\theta + B_1) + A_2k \cos(k\theta + B_2) + \cdots + A_pk^{p-1} \cos(k\theta + B_p)].$$

The sum of the solutions obtained in this way contains n arbitrary constants and is the general solution of the homogeneous equation (3.88). This generalizes Theorems 3.5 and 3.7.

The method of undetermined coefficients, as outlined in Section 3.4, may be used to obtain particular solutions of the complete equation (3.87). Theorem 3.6 (extended to the nth-order case) shows that the general solution of (3.87) is obtained as the sum of the general solution of the homogeneous equation and any particular solution of the complete equation. We know (Theorem 2.1) that the specification of the n initial values $y_0, y_1, \cdots, y_{n-1}$ allows a determination of the n arbitrary constants in this general solution and thus yields a unique solution of (3.87) satisfying the prescribed conditions.

As in Section 3.5, the limiting behavior of solutions depends upon both the initial values and on the roots of the auxiliary equation. Setting aside those exceptional initial values for which one or more of the constants in the general solution vanish, the limiting behavior of the solution of the homogeneous equation is determined by that part of the solution due to the root of the auxiliary equation which is largest in absolute value. If this maximum absolute value is denoted by ρ, as in Theorem 3.8, then *the solution of the homogeneous equation converges to* 0, *independently of the prescribed initial conditions, if and only if $\rho < 1$.* If $\rho > 1$, one generally obtains divergent solution sequences. A negative or complex root always contributes an oscillatory component to the solution and if ρ corresponds to such a root, then the limiting behavior of the solution is oscillatory, being damped or infinite as ρ is less or greater than 1.

In most cases, the roots of the auxiliary equation (3.89) are extremely difficult to calculate. Alternate methods of analysis must then be employed which enable conclusions to be made concerning the magnitude and nature of these roots without their explicit values being known. Some of these methods are discussed in Section 4.1.

Examples

(*a*) The third-order difference equation

$$y_{k+3} - 9y_{k+2} + 26y_{k+1} - 24y_k = 3$$

has the auxiliary equation

$$m^3 - 9m^2 + 26m - 24 = 0.$$

This equation can be factored into

$$(m - 2)(m - 3)(m - 4) = 0$$

so that the three roots are 2, 3, and 4.

The general solution of the homogeneous equation is given by

$$Y_k = C_1 2^k + C_2 3^k + C_3 4^k.$$

We try a constant function as a trial solution for the complete equation. If $y_k{}^* = C$ is to satisfy the equation,

$$C - 9C + 26C - 24C = 3$$

or $C = -\frac{1}{2}$. Hence the general solution of the complete difference equation is given by

$$y_k = C_1 2^k + C_2 3^k + C_3 4^k - \tfrac{1}{2}.$$

Only if the initial conditions are $y_0 = y_1 = y_2 = -\frac{1}{2}$ do all three constants C_1, C_2, and C_3 become zero. In all other cases, the solution sequence will diverge to $+\infty$ or $-\infty$, depending upon whether the coefficient of the dominating term is positive or negative.

(*b*) The difference equation[13]

(3.90) $$Y_{t+1} = 1.35 Y_t - 0.35 Y_{t-3}$$

can be rewritten in the standard form

(3.91) $$Y_{t+4} - 1.35 Y_{t+3} + 0.35 Y_t = 0.$$

The auxiliary equation

(3.92) $$m^4 - 1.35m^3 + 0.35 = 0$$

is easily seen to have $m = 1$ as a root. It therefore has $(m - 1)$ as a factor and we perform a long division[14] (of $m^4 - 1.35m^3 + 0.35$ by $m - 1$) to find that the auxiliary equation becomes

$$(m - 1)[m^3 - 0.35(m^2 + m + 1)] = 0.$$

We may detect one other real root by noting that the bracketed factor is negative when $m = 1$ and positive when $m = 2$. Hence a root exists between 1 and 2. The quantity in brackets is also positive when $m = 1.1$, so that the root is actually between 1 and 1.1. In this way, the root may be pinned down to any desired degree of accuracy. (Neisser finds it to be approximately 1.025.)

---

[13] This equation is considered in H. Neisser, "Depreciation, Replacement and Regular Growth," *Economic Journal*, 65 (1955), 159–161.

[14] The reader familiar with the process of *synthetic division* will find it a time-saving device here and in 1 of Problems 3.7.

That $m_1 = 1$ and $m_2 = 1.025$ are the only *real* roots of the auxiliary equation (3.92) follows from Descartes' rule of signs.[15] For $f(m) = m^4 - 1.35m^3 + 0.35$ has two changes of sign so there are at most these two *positive* roots. And $f(-m) = m^4 + 1.35m^3 + 0.35$ has no changes of sign so there are no negative roots. And of course, zero is not a root.

The auxiliary equation, being of degree 4, has four roots, two of which we know to be real. The other two are complex numbers and are found as follows. The bracketed quantity may be divided by its factor $(m - 1.025)$. The resulting quadratic factor (the reader should check this long division) is $m^2 + 0.675m + 0.342$. Hence the two remaining roots of (3.92) are the roots of the quadratic equation

$$m^2 + 0.675m + 0.342 = 0.$$

(Actually, this is only approximate due to the fact that m_2 is not exactly 1.025.) The quadratic formula yields the roots

$$m_3 = -0.34 + i\sqrt{0.23} \qquad m_4 = -0.34 - i\sqrt{0.23}.$$

The modulus of these roots is found, using (3.45), to be approximately $r = 0.6$. The general solution of (3.91) is therefore

$$Y_t = C_1 + C_2(1.025)^t + A(0.6)^t \cos(t\theta + B).$$

If $C_2 \neq 0$, the term $C_2(1.025)^t$ predominates and the sequence $\{Y_t\}$ diverges. The cosine term creates damped cyclical variations which are superimposed on the dominant explosive term. But if initial values are chosen for which $C_2 = 0$, then the constant term C_1 predominates and Y undergoes (convergent) damped oscillations about its limiting value C_1.

PROBLEMS 3.7

1. The following difference equations have been constructed so that their auxiliary equations have at least one integral root and hence can easily be factored. Find the remaining roots and write the general solution of each difference equation,

(a) $y_{k+3} - 6y_{k+2} + 11y_{k+1} - 6y_k = 0$

(b) $2y_{k+3} - 9y_{k+2} + 12y_{k+1} - 4y_k = 3$

(c) $y_{k+3} - 3y_{k+2} + 4y_{k+1} - 2y_k = 0$

(d) $y_{k+4} - 2y_{k+3} + 2y_{k+2} - 2y_{k+1} + y_k = 0$

(e) $y_{k+4} - 4y_{k+3} + 6y_{k+2} - 4y_{k+1} + y_k = 0$

---

[15] Let $f(m) = c_0 + c_1m + c_2m^2 + \cdots + c_nm^n$, where the c's are constants and $c_0 \neq 0$. A *change of sign* is said to occur at any particular term if that term has the opposite sign as the preceding, terms with zero coefficient not being counted. For example, $f(m) = m^6 - 2m^5 + 3m^3 + m^2 - m + 2$ has four changes of sign, occurring at the terms $-2m^5$, $+3m^3$, $-m$, and $+2$. If $f(-m)$ denotes the value of f at $-m$, then $f(-m) = m^6 + 2m^5 - 3m^3 + m^2 + m + 2$, so $f(-m)$ has two changes of sign. *Descartes' rule of signs: The equation $f(m) = 0$ cannot have more positive roots than $f(m)$ has changes of sign, or more negative roots than $f(-m)$ has changes of sign.*

A proof may be found on pp. 76–80 of Dickson, cited in footnote 12.

2. Consider the difference equation[16]

$$Y_{t+n+1} - (1 + A)Y_{t+n} + A Y_t = 0,$$

where A is a constant greater than $1/n$. Show that $m = 1$ is a root of the auxiliary equation and thus obtain the auxiliary equation in the form

$$(m - 1)[m^n - A(m^{n-1} + m^{n-2} + \cdots + m + 1)] = 0.$$

Use Descartes' rule of signs to show that the auxiliary equation can have at most one positive root in addition to $m = 1$. Show that the bracketed quantity is negative when $m = 1$ and positive for some sufficiently large value of m and conclude that this additional root exists and is greater than 1. What then is the nature of the solution sequence $\{Y_t\}$?

*3.8 LINEAR DIFFERENTIAL EQUATIONS WITH CONSTANT COEFFICIENTS

It is worth remarking that the theory of linear nth-order *differential* equations with constant coefficients is entirely analogous to the corresponding theory of difference equations. To illustrate this point, we briefly indicate the main results in the case $n = 2$. Detailed statements and proofs may be found in most textbooks on differential equations.

The linear second-order differential equation with constant coefficients can be written in the form

(3.93)
$$\frac{d^2y}{dx^2} + a_1 \frac{dy}{dx} + a_2 y = r(x),$$

where y and r are functions defined for real numbers x in some interval $a \le x \le b$, and a_1 and a_2 are constants, with $a_2 \ne 0$. Suppose the value of y and its first derivative are specified at a single point of this interval. Then an existence and uniqueness theorem shows that there is one and only one solution of (3.93) satisfying these two initial conditions.

The general solution of (3.93) is the sum of the general solution of the homogeneous equation

$$\frac{d^2y}{dx^2} + a_1 \frac{dy}{dx} + a_2 y = 0$$

and any particular solution of the complete equation (3.93). The general solution of the homogeneous equation is given by $C_1 y^{(1)} + C_2 y^{(2)}$ pro-

---

[16] Equation (3.91) is the special case for which $n = 3$ and $A = 0.35$. For the economic significance of the general equation of order n, see the reference cited in footnote 13 as well as E. D. Domar, "Depreciation, Replacement and Growth," *Economic Journal*, *63* (1953).

vided that the two solutions $y^{(1)}$ and $y^{(2)}$ form a fundamental set of solutions in the interval $a \leq x \leq b$. Fundamental sets may be defined by the requirement that a certain determinant (the Wronskian) be different from zero.

Such fundamental sets are easy to find and one can prove the following key result in the theory (cf. Theorem 3.7):

THEOREM. *If*

$$(3.94) \qquad \frac{d^2y}{dx^2} + a_1 \frac{dy}{dx} + a_2 y = 0,$$

where a_1 and a_2 are constants and $a_2 \neq 0$, and if m_1 and m_2 are the two roots of the auxiliary equation

$$m^2 + a_1 m + a_2 = 0,$$

the general solution of the differential equation (3.94) *is given by*

$$C_1 e^{m_1 x} + C_2 e^{m_2 x}$$

if m_1 and m_2 are real and unequal; by

$$(C_1 + C_2 x) e^{m_1 x}$$

if m_2 is equal to m_1; and by

$$A e^{ax} \cos (bx + B)$$

if m_1 and m_2 are complex conjugates $a \pm bi$.

To this general solution must be added a particular solution of the complete equation. The method of undetermined coefficients, as well as other techniques, is available for finding such particular solutions.

The theory of the general nth-order linear differential equation with constant coefficients is a straightforward generalization of the case $n = 2$ just considered.

The extraordinary similarity of this theory of differential equations with the corresponding theory of difference equations is a consequence of the close connections, discussed in Sections 1.7 and 2.12, between the fundamental operations of the difference and differential calculus.

We conclude by referring the advanced reader to a treatise[17] on mixed difference-differential equations.

[17] Edmund Pinney, *Ordinary Difference-Differential Equations*, Univ. of California Press, Berkeley, 1958. An equation studied by Tinbergen in a theory of cycles in the shipbuilding industry is considered on p. 188.

Selected
Topics

4.1 EQUILIBRIUM AND STABILITY

As noted and illustrated in Section 3.5, the nature of the solution of a linear difference equation with constant coefficients, especially its limiting behavior, is dependent upon both the initial values prescribed for the solution and the roots of the auxiliary equation. However, for the second-order homogeneous equation

$$(4.1) \qquad y_{k+2} + a_1 y_{k+1} + a_2 y_k = 0$$

we were able, in Theorem 3.8, to formulate a necessary and sufficient condition for the solution to converge to 0 *independently of the initial values* y_0 and y_1. This condition requires that both the roots of the auxiliary equation

$$(4.2) \qquad m^2 + a_1 m + a_2 = 0$$

be less than 1 in absolute value. Theorem 3.8 is a key result in the discussion of the *stability* of a difference equation, to which we now turn.

Consider the complete difference equation

$$(4.3) \qquad y_{k+2} + a_1 y_{k+1} + a_2 y_k = r$$

where we write r in place of the usual r_k since we now assume a constant ($= r$) right-hand term.

If (4.3) has a constant function as solution, then the value of this function is called an *equilibrium* (or *stationary*) *value* of y. Putting $y_k = y^*$, a constant, in (4.3), we find

$$y^* + a_1 y^* + a_2 y^* = r.$$

Hence, if $1 + a_1 + a_2 \neq 0$, then

$$(4.4) \qquad y^* = \frac{r}{1 + a_1 + a_2}$$

is an equilibrium value of y.   Such an equilibrium value has the following property: if two consecutive values of any solution y of (4.3) are equal to y^*, then all succeeding values of y are equal to y^*.   This is an immediate consequence of the fact that if y_k and y_{k+1} are put equal to y^* in (4.3), then $y_{k+2} = y^*$ also.

The equilibrium value is said to be stable[1] (or the difference equation (4.3) is stable) if every solution of (4.3), independently of the prescribed initial conditions y_0 and y_1, converges to y^*, i.e., if

$$(4.5) \qquad \lim_{k \to \infty} y_k = y^* \qquad (\textit{for all } y_0 \textit{ and } y_1).$$

Since a displacement from the equilibrium value is equivalent to considering a new solution with different initial conditions, we may alternatively define a stable equilibrium as one for which *any* displacement from equilibrium is followed by a sequence of values of y which again converge to this equilibrium.

It is convenient to define a new function z which measures the deviation of a solution y of (4.3) from its equilibrium value y^*;   i.e., we let

$$(4.6) \qquad z_k = y_k - y^*.$$

Then

$$\begin{aligned} z_{k+2} + a_1 z_{k+1} + a_2 z_k &= (y_{k+2} + a_1 y_{k+1} + a_2 y_k) - (1 + a_1 + a_2)y^* \\ &= y_{k+2} + a_1 y_{k+1} + a_2 y_k - r \\ &= 0 \end{aligned}$$

since y is a solution of (4.3).   Hence z is a solution of the homogeneous difference equation (4.1).   The definition of stability requires that z,

---

[1] A number of different definitions of stability appear in the literature.   We here define what is known as "perfect stability of the first kind."   See P. A. Samuelson, *Foundations of Economic Analysis*, Harvard Univ. Press, Cambridge, 1948, p. 261.   For a nonmathematical discussion of various kinds of stability, see H. Leibenstein, *A Theory of Economic-Demographic Development*, Princeton Univ. Press, Princeton, 1954, Chap. 3.   (In this book difference equations are used to study the relationships between factors causing economic growth and those affecting population size.)   A related concept of stability occurs in numerical analysis, where a stable numerical procedure is one for which an error, once made, is damped as the calculations proceed.   See P. D. Lax and R. D. Richtmyer, "Survey of the Stability of Linear Finite Difference Equations," *Communications on Pure and Applied Mathematics*, 9 (1956), 267–293. Stability of differential equations is discussed in R. Bellman, *Stability Theory of Differential Equations*, McGraw-Hill, New York, 1953.

the deviation of y from its equilibrium value, converge to 0 for every pair of initial values z_0 and z_1. Since z is a solution of (4.1), Theorem 3.8 supplies an immediate proof of the following result.

THEOREM 4.1.   *A necessary and sufficient condition for the equilibrium value y^* in* (4.4) *to be stable is $\rho < 1$, where*

$$\rho = \max(|m_1|, |m_2|)$$

and m_1 and m_2 are the roots of the auxiliary equation (4.2).

Because of this theorem it becomes important to learn what restrictions should be placed on coefficients a_1 and a_2 in the auxiliary equation so that ρ will indeed be less than 1.   By the quadratic formula, the two roots of (4.2) are

$$(4.7) \qquad \frac{-a_1 \pm \sqrt{a_1{}^2 - 4a_2}}{2}.$$

The quantity $a_1{}^2 - 4a_2$ appearing under the square root sign is the *discriminant* of the auxiliary equation.   The two roots m_1 and m_2 are real and unequal, real and equal, or complex conjugates, according as the discriminant is positive, zero, or negative.

Necessary conditions for $\rho < 1$ may be obtained by *assuming* $\rho < 1$ and determining the resulting restrictions placed on coefficients a_1 and a_2. We do this by considering separately the cases of (1) real and (2) complex roots.

CASE 1.   REAL ROOTS

Since $\rho < 1$, both roots are between -1 and 1.   Hence, from (4.7) we have the two inequalities,

$$-2 < -a_1 + \sqrt{a_1{}^2 - 4a_2} < 2 \qquad -2 < -a_1 - \sqrt{a_1{}^2 - 4a_2} < 2,$$

or, adding a_1 throughout,

$$(4.8) \qquad -2 + a_1 < \sqrt{a_1{}^2 - 4a_2} < 2 + a_1,$$

$$(4.9) \qquad -2 + a_1 < -\sqrt{a_1{}^2 - 4a_2} < 2 + a_1.$$

The sum of the roots is $-a_1$ and since we are assuming each root is between -1 and 1, we certainly have $-2 < -a_1 < 2$ or $-2 + a_1 < 0$ and $2 + a_1 > 0$.   Thus the first inequality in (4.8) and the second in (4.9) are always true and hence yield no restrictions on a_1 and a_2.   The second inequality in (4.8), when squared, reads

$$a_1{}^2 - 4a_2 < (2 + a_1)^2$$

or

$$a_1{}^2 - 4a_2 < 4 + 4a_1 + a_1{}^2,$$

from which

(4.10) $1 + a_1 + a_2 > 0.$

The first inequality in (4.9), when squared,[2] becomes

$$(-2 + a_1)^2 > a_1{}^2 - 4a_2$$

or, upon simplification,

(4.11) $1 - a_1 + a_2 > 0.$

Conditions (4.10) and (4.11) are necessary if both real roots of the auxiliary equation are to be less than 1 in absolute value.

CASE 2.   COMPLEX CONJUGATE ROOTS

Here a necessary condition is easy to obtain.   For if the two roots are $m_1 = a + bi$ and $m_2 = a - bi$, the product of these roots is $m_1 m_2 = a^2 + b^2$ and this, by (3.45), is just r^2, the square of the modulus (or absolute value) of each complex root.   We are assuming $r < 1$.   Hence $r^2 = m_1 m_2 < 1$. But $m_1 m_2$, the product of the roots of (4.2), is equal to a_2.   Hence it is necessary that $a_2 < 1$ or

(4.12) $1 - a_2 > 0.$

The three necessary conditions (4.10), (4.11), and (4.12) turn out also to be sufficient for $\rho < 1$.   This may be verified by showing, using these inequalities, that it is possible to reverse the steps in the above calculations and arrive at $\rho < 1$ in each case.   In this way, one completes the proof of the following result.

THEOREM 4.2.   *The conditions*

(4.13) $1 + a_1 + a_2 > 0$ $1 - a_1 + a_2 > 0$ $1 - a_2 > 0$

are necessary and sufficient for both roots of

$$m^2 + a_1 m + a_2 = 0$$

to be less than 1 in absolute value.

In view of Theorem 4.1, it follows that *the three inequalities in* (4.13) *are necessary and sufficient conditions for the stability of the equilibrium value* y^*.

---

[2] Squaring inequalities requires care: If $a < b$ and a and b are both *positive*, then $a^2 < b^2$, but if $a < b$ and a and b are both *negative*, then $a^2 > b^2$. Thus $3 < 4$ implies $3^2 < 4^2$, but $-3 < -2$ implies $(-3)^2 > (-2)^2$.

Example

Return to the Hansen-Samuelson difference equation (3.65) which we repeat here:

(4.14) $$Y_{t+2} - \alpha(1 + \beta) Y_{t+1} + \alpha\beta Y_t = 1.$$

Recall that Y_t denotes the national income in period t and α and β are the *marginal propensity to consume* and the *relation* respectively. We assume $\alpha > 0$ and $\beta > 0$.

If Y^* is a constant solution of (4.14),

$$Y^* - \alpha(1 + \beta) Y^* + \alpha\beta Y^* = 1$$

from which we find the equilibrium value of national income

(4.15) $$Y^* = \frac{1}{1 - \alpha} \qquad \alpha \neq 1.$$

The stability conditions (4.13) when applied to the difference equation (4.14) become

$$1 - \alpha(1 + \beta) + \alpha\beta > 0 \qquad 1 + \alpha(1 + \beta) + \alpha\beta > 0 \qquad 1 - \alpha\beta > 0.$$

The second of these is satisfied automatically since α and β are both positive. The first and third may be rewritten as

(4.16) $$\alpha < 1 \qquad \text{and} \qquad \alpha\beta < 1.$$

These two conditions are the necessary and sufficient conditions for the income Y^* to be a stable equilibrium value: *both* the marginal propensity to consume and its product with the relation must be less than 1. If these requirements are fulfilled, the sequence of income values will converge to Y^* independently of the initial conditions prescribed.

This convergence to Y^* will be oscillatory if the roots are complex numbers. For this, the discriminant $a_1{}^2 - 4a_2$ must be negative. In the present case,

$$\alpha^2(1 + \beta)^2 - 4\alpha\beta < 0$$

or

(4.17) $$\alpha < \frac{4\beta}{(1 + \beta)^2}.$$

The roots of the auxiliary equation are complex if (4.17) is satisfied and real otherwise. These calculations can be conveniently summarized graphically, but for this we refer the reader to the Samuelson article cited on p. 6.

Similar definitions of equilibrium value and its stability apply to difference equations of order 1 (cf. 1 of Problems 4.1) and order n with $n > 2$. In the general case, conditions are known[3] for which the auxiliary

---

[3] P. A. Samuelson, "Conditions that the roots of a polynomial be less than unity in absolute value," *Annals of Mathematical Statistics*, *12* (1941), 360–364; J. S. Chipman, "The Multi-Sector Multiplier," *Econometrica*, *18* (1950), 355–374. For examples of the application of these conditions to difference equation models of income determination, see R. S. Eckaus, "The Stability of Dynamic Models," *Review of Economics and Statistics*, *39* (1957), 172–182.

equation, which is now the polynomial of nth degree in (3.89), has all its roots less than 1 in absolute value. As in the case $n = 2$, this is necessary and sufficient for stability.

Return to the difference equation (4.3) but now with an arbitrary right-hand member rather than a constant:

$$(4.18) \qquad y_{k+2} + a_1 y_{k+1} + a_2 y_k = r_k.$$

A particular solution, y^*, of this complete equation will not be a constant function, but will vary with k. The general solution of (4.18) is given by

$$y_k = Y_k + y_k^*,$$

where Y is the solution of the corresponding homogeneous equation. If the roots of the auxiliary equation are both less than 1 in absolute value, we know the sequence $\{Y_k\}$ will converge to 0. When this happens, the remaining part of the solution is called a *moving equilibrium*, the solution y approaching (as k increases) this equilibrium function y^* as the influence of Y diminishes.

Example[4]

Make the following assumptions relating income (y), consumption (C), and investment (i):

(i) Income equals consumption plus investment, all for period n; i.e.,

$$y_n = C_n + i_n.$$

(ii) Consumption in any period is a linear function of the incomes of the two preceding periods; i.e.,

$$C_n = c_1 y_{n-1} + c_2 y_{n-2} + K,$$

where c_1, c_2, and K are constants.

(iii) Investment increases a fixed amount h each period; i.e.,

$$i_{n+1} = i_n + h \qquad \text{or} \qquad i_n = i_0 + nh.$$

Combining these relations, the following difference equation is obtained for the income function:

$$y_n = c_1 y_{n-1} + c_2 y_{n-2} + K + i_0 + nh.$$

We write this in the equivalent form

$$(4.19) \qquad y_{n+2} - c_1 y_{n+1} - c_2 y_n = nh + A,$$

where, for simplicity, we have combined all constants into $A \ (= K + i_0 + 2h)$.

Consider the special case $c_1 = \frac{1}{2}$, $c_2 = \frac{1}{4}$. (The general case is taken up in 5 of Problems 4.1.) Now

$$(4.20) \qquad y_{n+2} - \tfrac{1}{2} y_{n+1} - \tfrac{1}{4} y_n = nh + A.$$

---

[4] J. R. Hicks, *A Contribution to the Theory of the Trade Cycle*, Oxford Univ. Press, London, 1950, pp. 174–175.

The auxiliary equation

$$m^2 - \tfrac{1}{2}m - \tfrac{1}{4} = 0$$

has coefficients which satisfy (4.13) and so the solution of the homogeneous equation converges to 0. This solution is actually given by

$$Y_n = C_1 m_1{}^n + C_2 m_2{}^n,$$

where $m_1 = (1 + \sqrt{5})/4$ and $m_2 = (1 - \sqrt{5})/4$.

A solution of the complete equation (4.20) can be found by the method of undetermined coefficients. A trial solution of the form

$$y_n{}^* = \alpha + \beta n,$$

if substituted into (4.20), is found to satisfy the equation if

$$\beta = 4h \qquad \alpha = 4(A - 6h).$$

The general solution of (4.20) is therefore

$$y_n = Y_n + y_n{}^*.$$

Since $Y_n \to 0$ as n increases, the income becomes increasingly close to its moving equilibrium, the linear function y^*. Thus, the income adjusts to the expansion caused by an investment function which, according to assumption (iii), is increasing linearly with time.

PROBLEMS 4.1

1. If

$$y_{k+1} = A y_k + B,$$

show that an equilibrium value of y is given by $B/(1 - A)$ if $A \neq 1$. Prove, using Theorem 2.7, that $-1 < A < 1$ is a necessary and sufficient condition for this equilibrium to be stable.

2. (a) Verify that the values $\alpha = 0.5$ and $\beta = 1$ satisfy both (4.16) and (4.17) and conclude that in this case the solution Y of (4.14) undergoes damped oscillations toward the stable equilibrium value $Y^* = 2$. (b) Verify that $\alpha = 0.8$ and $\beta = 2$ do not satisfy the stability conditions (4.16) but do satisfy (4.17). Conclude that Y now undergoes infinite oscillations. (Review the discussion in Example 1 of Section 3.6.)

3. If the second-order difference equation is written in the form

$$y_{k+2} - B y_{k+1} + C y_k = 0$$

with B and C constants, show[5] that the stability conditions (4.13) may be written

$$-1 - C < B < 1 + C \qquad C < 1.$$

----

[5] For an economic application, see W. S. Vickrey, "Stability Through Inflation," Chap. 4 in K. K. Kurihara (ed.), *Post Keynsian Economics*, Rutgers Univ. Press, New Brunswick, 1954.

4. Consider the difference equation[6]

$$(1 - \epsilon_2 + \beta)Y_{k+2} - [\epsilon_1\gamma - \epsilon_2\gamma + \alpha + \beta(1 + \gamma)]Y_{k+1} + \beta\gamma Y_k$$
$$= \eta(\epsilon_1 - \epsilon_2) + \delta_1 + \delta_2$$

where all coefficients are positive constants with

$$1 > \epsilon_1 > \epsilon_2 \qquad \alpha < 1 - \epsilon_1\gamma - \epsilon_2(1 - \gamma) \qquad \gamma < 1.$$

Find the equilibrium value of Y and apply the conditions (4.13) to show that the equilibrium is stable.

5. In (4.19) assume $|c_1| + |c_2| < 1$. (a) Show, by verifying conditions (4.13), that the solution of the corresponding homogeneous equation converges to 0. (b) Show that the moving equilibrium is of the form $\alpha + \beta n$.

6. In (4.19) suppose $c_2 = 0$ (i.e., consumption depends only on income one period before.) Solve the difference equation and show that $y_n = C(c_1)^n + \alpha + \beta n$ and hence deduce that the moving equilibrium is of the form $\alpha + \beta n$ if $|c_1| < 1$.

4.2 FIRST-ORDER EQUATIONS AND COBWEB CYCLES

No general method exists for solving linear difference equations of order greater than 1 if the coefficients are not constants. But as we now show, it is possible to solve the first-order linear equation

$$(4.21) \qquad y_{k+1} - a_k y_k = r_k \qquad k = 0, 1, 2, \cdots,$$

where a and r are functions defined for the indicated set of k-values. We assume $a_k \neq 0$ for all k.

By the existence and uniqueness theorem of Section 2.3, there is exactly one solution of (4.21) for which the value y_0 can be prescribed.

The homogeneous equation

$$(4.22) \qquad y_{k+1} - a_k y_k = 0$$

allows a stepwise calculation of a solution. For we find

$$y_1 = a_0 y_0, \qquad y_2 = a_1 y_1 = a_0 a_1 y_0, \qquad y_3 = a_2 y_2 = a_0 a_1 a_2 y_0$$

and, in general, if

$$(4.23) \quad y_0 = C, \qquad y_k = C(a_0 a_1 a_2 \cdots a_{k-1}) \qquad k = 1, 2, 3, \cdots,$$

then y is a solution of (4.22). Since y_0 is arbitrary, we write it as a constant C. Particular solutions of the homogeneous equation are found for each choice of this constant.

---

[6] A. Smithies, "The Behavior of Money National Income under Inflationary Conditions," *Quarterly Journal of Economics*, 57 (1942), 113–128; reprinted in American Economic Association, *Readings in Fiscal Policy*, Richard D. Irwin, Homewood, Ill., 1955.

Just as a sum of terms is written more easily using the $\sum$ notation, so a product is written using the symbol $\prod$.   The following examples illustrate this notation:

$$1 \cdot 2 \cdot 3 = \prod_{j=1}^{3} j \qquad \frac{1}{2} \cdot \frac{2}{3} \cdot \frac{3}{4} \cdot \frac{4}{5} = \prod_{j=1}^{4} \frac{j}{j+1}$$

$$1^2 \cdot 2^2 \cdot 3^2 = \prod_{j=1}^{3} j^2 \qquad x(x-1)(x-2) = \prod_{j=0}^{2} (x-j)$$

$$a_0 a_1 a_2 a_3 = \prod_{j=0}^{3} a_j \qquad a_0 a_1 a_2 \cdots a_9 = \prod_{j=0}^{9} a_j$$

With this notation, (4.23) becomes

(4.24) $$y_0 = C, \qquad y_k = C \prod_{j=0}^{k-1} a_j \qquad k = 1, 2, 3, \cdots.$$

Now let y denote a solution of the complete equation and suppose y may be written in the form

(4.25) $$y_k = u_k v_k$$

as the product of two functions u and v to be determined.   Then

$$
\begin{aligned}
y_{k+1} - a_k y_k &= u_{k+1} v_{k+1} - a_k u_k v_k \\
&= u_{k+1}(v_k + \Delta v_k) - a_k u_k v_k \\
&= v_k(u_{k+1} - a_k u_k) + u_{k+1}\Delta v_k.
\end{aligned}
$$

From this calculation it follows if we choose the functions u and v so that

(4.26) $$u_{k+1} - a_k u_k = 0$$

and

(4.27) $$u_{k+1}\Delta v_k = r_k,$$

then y is indeed a solution of the complete equation (4.21).   The function u is a solution of the homogeneous equation (4.26) and we have already seen that such a solution is given by

(4.28) $$u_0 = C, \qquad u_k = C \prod_{j=0}^{k-1} a_j \qquad k = 1, 2, 3, \cdots$$

for any value of C. Since $a_j \neq 0$ for all j, we obtain a solution u for which $u_k \neq 0$ for all k by choosing $C \neq 0$.

For such a solution we can divide (4.27) by u_{k+1} to obtain

$$\Delta v_k = \frac{r_k}{u_{k+1}}.$$

Since r_k and u_{k+1} are known, the first difference of v is a known function and v is an indefinite sum of this function. This indefinite sum is determined up to an additive constant. With the notation of Section 1.6, with c a constant,

$$(4.29) \qquad v_k = \Delta^{-1} \frac{r_k}{u_{k+1}} + c.$$

To summarize: find u as a nonzero solution of the homogeneous equation (4.22); such a solution is given by (4.28) with $C \neq 0$. Find v as the indefinite sum in (4.29) containing the arbitrary constant c. Then the product of u and v is a solution of the first-order equation (4.21) for every value of c. It is the general solution of (4.21), for if a particular solution with y_0 specified is required, this initial condition may be used to determine the value of c.

Examples

(*a*) It is instructive to first apply this procedure to the difference equation

$$y_{k+1} = Ay_k + B,$$

where A and B are constants. Here $a_k = A$, $r_k = B$. We choose $C = 1$ in (4.28) to find

$$u_k = \prod_{j=0}^{k-1} A = A^k \qquad k = 0, 1, 2, \cdots.$$

For v we use (4.29) and obtain

$$\Delta^{-1} \frac{B}{A^{k+1}} = \frac{B}{A} \Delta^{-1} \left(\frac{1}{A} \right)^k.$$

But, as may be checked by taking differences of both sides, or by applying the result of Problem 3, Section 1.6,

$$\Delta^{-1} \left(\frac{1}{A} \right)^k = \begin{cases} \dfrac{A}{1-A} \left(\dfrac{1}{A} \right)^k & \text{if } A \neq 1 \\ k & \text{if } A = 1. \end{cases}$$

Hence

$$v_k = \begin{cases} \dfrac{B}{1-A} \left(\dfrac{1}{A} \right)^k + c & \text{if } A \neq 1 \\ Bk + c & \text{if } A = 1 \end{cases}$$

and the solution y is given by

$$y_k = u_k v_k = \begin{cases} cA^k + \dfrac{B}{1-A} & \text{if } A \neq 1 \\ c + Bk & \text{if } A = 1. \end{cases}$$

It is easy to find c in terms of y_0. This solution is then identical to that obtained in Section 2.4.

(*b*) For the difference equation

$$y_{k+1} - (k + 1)y_k = (k + 1)!$$

we find

$$u_k = \prod_{j=0}^{k-1} (j + 1) = 1 \cdot 2 \cdot 3 \cdots k = k!$$

is a particular solution of the homogeneous equation. Then, by (4.29),

$$v_k = \Delta^{-1} \frac{(k + 1)!}{(k + 1)!} + c = \Delta^{-1} 1 + c = k + c.$$

Hence the solution is given by

$$y_k = k!(k + c).$$

If $y_0 = 2$ is prescribed, then $(0! = 1)$

$$y_0 = c = 2$$

and the required solution is

$$y_k = k!(k + 2).$$

The so-called "cobweb" phenomena[7] in economics afford excellent examples in which simple first-order difference equations may be applied. For definiteness, consider the following kind of situation. Farmers decide on the basis of this year's price for a certain commodity the acreage they will plant with that crop. Anticipating that the price level will be maintained, if the price is high one year, farmers tend to plant heavily. The following year, when the crop is harvested and brought to market, the supply exceeds the demand, prices fall, and farmers cut acreage devoted to this particular commodity. When the next crop is harvested, supply may be below demand (we assume no inventory carry-over), prices increase, farmers plant more, next year's crop exceeds demand, prices fall, etc.

For a mathematical analysis, we study three functions defined for $t = 0, 1, 2, \cdots$ as follows: $S_t =$ number of units supplied in period t, $D_t =$ number of units demanded in period t, $p_t =$ price per unit in period t. We make the following assumptions:

(i) A price-demand relation is specified in which quantity demanded is determined by the price at the time of purchase; i.e., a function f exists for which $D_t = f(p_t)$.

(ii) A price-supply curve relates the supply in any period with the price one period before; i.e., a function g exists for which $S_{t+1} = g(p_t)$.

(iii) The market price is determined by the available supply, transactions occurring at the price at which the quantity demanded and the quantity

[7] M. Ezekiel, "The Cobweb Theorem," *Quarterly Journal of Economics*, 52 (1938), 255–280; reprinted in *Readings in Business Cycle Theory*, Blakiston Co., Philadelphia, 1944. N. S. Buchanan, "A Reconsideration of the Cobweb Theorem," *Journal of Political Economy*, 47 (1939), 67–81; reprinted in R. V. Clemence (ed.), *Readings in Economic Analysis*, Vol. 1, Addison-Wesley, Cambridge, Mass., 1950.

supplied are equal; i.e., p_t is determined as the solution of the equation $S_t = D_t$.

Now suppose p_0 is prescribed. If the functions f and g are known, we may calculate S_1 from (ii) and D_1 from (iii) and so obtain p_1 from the price-demand curve in (i). The process may be repeated (starting with p_1) to obtain p_2, etc. The sequence of prices $\{p_t\} = p_0, p_1, p_2, \cdots$ is to be studied, special interest attaching to the possibility of oscillatory behavior.

To proceed further we assume f and g are linear functions or, equivalently, the *price-demand and price-supply curves are straight lines.*[8] The equations in (i) and (ii) are thereby simplified to

$$(4.30) \qquad D_t = -m_d p_t + b_d \qquad m_d > 0, \, b_d > 0$$

$$(4.31) \qquad S_{t+1} = m_s p_t + b_s \qquad m_s > 0, \, b_s > 0$$

where $-m_d$ and b_d are the slope and D-intercept for the demand curve, with m_s and b_s the corresponding constants for the supply curve. The slope of the demand curve is taken to be negative, that of the supply curve positive: an *increase* of 1 unit in price produces a *decrease* of m_d units in demand, but an *increase* of m_s units in supply. At price 0, demand and supply are b_d and b_s respectively.

By hypothesis (iii), price is determined by the equality of supply and demand, or writing this for period $t + 1$,

$$(4.32) \qquad S_{t+1} = D_{t+1}.$$

Hence,

$$m_s p_t + b_s = -m_d p_{t+1} + b_d$$

or

$$(4.33) \qquad p_{t+1} = A p_t + B$$

where

$$A = -\frac{m_s}{m_d} \qquad B = \frac{b_d - b_s}{m_d}.$$

---

[8] This restriction is made only to simplify the mathematical analysis. Nonlinear supply and demand curves were first studied by W. Leontief in "Verzogerte Angebotsanpassung und Partielles Gleichgewicht," *Zeitschrift fur Nationalokonomie*, 5 (1934), 670–676. An example can be found in P. A. Samuelson, "Dynamic Process Analysis," in H. S. Ellis (ed.), *A Survey of Contemporary Economics*, Blakiston Co., Philadelphia, 1948. For the consequences of allowing time-dependent supply and demand schedules as well as of permitting suppliers to learn from experience (instead of constantly being disappointed in their expectation that the price level will remain the same from one period to the next), see R. M. Goodwin, "Dynamical Coupling With Especial Reference to Markets Having Production Lags," *Econometrica*, 15 (1947), 181–204.

The first-order difference equation (4.33) was studied in Section 2.4. Since A is negative, it follows from Theorem 2.7 that the sequence $\{p_t\}$ is always oscillatory, being damped or convergent [with limit $p^* = B/(1 - A)$] if $-1 < A < 0$, undergoing finite oscillations if $A = -1$, and infinite oscillations if $A < -1$. Since A is the ratio of the slopes of the supply

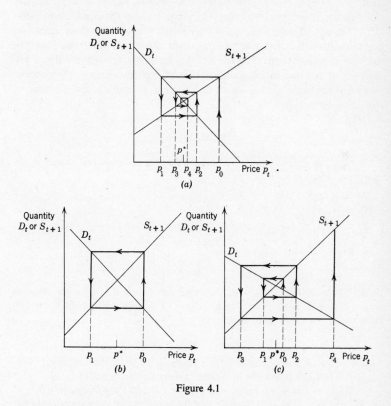

Figure 4.1

and demand curves, this ratio is seen to determine the behavior of the price sequence.

The reason for calling this "cobweb" behavior becomes clear with a graphical analysis. In Figure 4.1 we illustrate the three cases $-1 < A < 0$, $A = -1$, and $A < -1$. Starting with p_0 we find S_1 by moving vertically to the supply line, then move horizontally (since $D_1 = S_1$) to find D_1, which determines the price p_1 on the price axis. The supply S_2 is found on the supply line directly above p_1, and again D_2 ($= S_2$) is found by moving horizontally to the demand line, etc. The intersection of the supply and demand lines determines the equilibrium price p^*. In Figure 4.1(a) we find a cobweb moving toward this equilibrium point, prices alternating above and below, but converging to, p^*. In Figure 4.1(b), the cobweb consists of one square endlessly repeated, prices oscillating finitely between just two values. In Figure 4.1(c), the cobweb moves away from the intersection point, prices oscillating infinitely about the equilibrium value p^*. The equilibrium p^* is *stable* only in the first of these three cases.

Without the linearity assumptions made above, the difference equation for the determination of price will be nonlinear. Such problems are difficult at best and in general incapable of explicit solution. We do not consider nonlinear difference equations, but mention a few sources of further information.[9]

PROBLEMS 4.2

1. Find the general solution of each difference equation and then find the particular solution for which $y_0 = 1$.

(a) $y_{k+1} - (k + 1)y_k = 0$
(b) $y_{k+1} - (k + 1)y_k = 2^k(k + 1)!$
(c) $y_{k+1} - 3^k y_k = 0$
(d) $y_{k+1} + (-1)^k y_k = 0$

2. In the difference equation

$$y_{k+1}^2 - 3y_{k+1}y_k + 2y_k^2 = 0$$

make the substitution $z_k = y_{k+1}/y_k$ and show that the equation becomes $(z_k - 2)(z_k - 1) = 0$, from which $z_k = 2$ or $z_k = 1$. Hence find the two solutions $y^{(1)}$ and $y^{(2)}$ given by

$$y_k^{(1)} = y_0 2^k \qquad \text{and} \qquad y_k^{(2)} = y_0.$$

These are two different solutions, both assuming the same initial value y_0. Why does this fact not violate the uniqueness theorem of Section 2.3?

3. If $y_{k+1} = 1/(1 - y_k)$ and $y_0 = a \neq 1$ is prescribed, show that $y_1 = 1/(1 - a)$, $y_2 = (a - 1)/a$, $y_3 = a$. Conclude that the sequence $\{y_k\}$ is an unending repetition of only three values.

---

[9] The difference equation $y_{k+1} = ay_k^2 + by_k + c$ is analyzed in T. W. Chaundy and E. Phillips, "The Convergence of Sequences Defined by Quadratic Recurrence-Formulae," *Quarterly Journal of Mathematics (Oxford Series)*, 7 (1936), 74–80. Graphical methods are applied to a nonlinear difference equation in M. S. Klamkin, "Geometric Convergence," *American Mathematical Monthly*, 60 (1953), 256–259. An analysis of the general difference equation of first order, especially with respect to stability properties, can be found in P. A. Samuelson, *Foundations of Economic Analysis*, Harvard Univ. Press, Cambridge, 1948, pp. 302–308. See also footnote 10.

4. Show that the difference equation[10]

$$y_{k+1} = \alpha - \frac{\beta}{y_k}$$

becomes the linear equation

$$z_{k+2} - \alpha z_{k+1} + \beta z_k = 0$$

if $y_k = z_{k+1}/z_k$. In the special case $\alpha = \beta = 4$, find z_k and show that

$$y_k = 2\,\frac{C_1 + C_2(k + 1)}{C_1 + C_2 k}.$$

Show that the sequence $\{y_k\}$ converges with limit 2.

5. (a) Suppose the slopes and intercepts in the price-demand and price-supply equations (4.30) and (4.31) are $m_d = b_d = 2$ and $m_s = b_s = 1$. Sketch a graph like Figure 4.1 and analyze the behavior of the price sequence. Find p^* and determine whether this equilibrium price is stable. (b) Same with $m_d = 1$, $b_d = 5$, $m_s = b_s = 1$. (c) Same with $m_d = 1$, $b_d = 7$, $m_s = 2$, $b_s = 1$.

6. The following second-order difference equation arises in finding the equilibrium distribution of individuals in various income ranges:[11]

$$x_n = \frac{0.3}{n} x_{n-1} + \left(0.9 - \frac{0.3}{n + 1}\right) x_n + 0.1 x_{n+1} \qquad n = 1, 2, 3, \cdots$$

with x_0 and $x_1 = 3x_0$ prescribed. This equation is linear but does not have constant coefficients. To find the solution proceed as follows:

(a) Show that algebraic simplification allows the equation to be written in the form

$$x_{n+1} - \frac{3}{n + 1} x_n = x_n - \frac{3}{n} x_{n-1} \qquad n = 1, 2, 3, \cdots.$$

(b) Let

$$y_n = x_n - \frac{3}{n} x_{n-1} \qquad n = 1, 2, 3, \cdots$$

and conclude that $y_{n+1} = y_n$ from which $y_n = C$, a constant. Use the prescribed initial conditions to show $y_1 = C = 0$ so that $y_n = 0$ or

$$x_n - \frac{3}{n} x_{n-1} = 0 \qquad n = 1, 2, 3, \cdots.$$

(c) Solve this first-order homogeneous difference equation to find

$$x_n = \frac{3^n}{n!} x_0 \qquad n = 0, 1, 2, \cdots.$$

---

[10] For a detailed analysis of this nonlinear equation see L. Brand, "A Sequence Defined by a Difference Equation," *American Mathematical Monthly*, 62 (1955), 489–492.

[11] D. G. Champernowne, "A Model of Income Distribution," *Economic Journal*, 63 (1953), 318–351.

7. Show that if

$$x_n = \alpha_{n-1}x_{n-1} + (1 - \alpha_n - \beta_n)x_n + \beta_{n+1}x_{n+1} \qquad n = 1, 2, 3, \cdots$$

with x_0 and $x_1 = (\alpha_0/\beta_1)x_0$ prescribed, and $\alpha_0, \alpha_1, \alpha_2, \cdots$ and $\beta_1, \beta_2, \beta_3, \cdots$ arbitrary numbers (no β being 0), then

$$x_n = x_0 \prod_{j=0}^{n-1} \left(\frac{\alpha_j}{\beta_{j+1}} \right)$$

(*Hint:* Write the equation in the form

$$\beta_{n+1}x_{n+1} - \alpha_n x_n = \beta_n x_n - \alpha_{n-1}x_{n-1} \qquad n = 1, 2, 3, \cdots$$

and follow the method of Problem 6 which is the special case given by

$$\alpha_n = \frac{0.3}{n + 1} \qquad \beta_{n+1} = 0.1 \qquad n = 0, 1, 2, \cdots.)$$

4.3 A CHARACTERISTIC-VALUE PROBLEM

Consider the homogeneous second-order difference equation

$$(4.34) \quad y_{k+2} - (2 - \alpha)y_{k+1} + y_k = 0 \qquad k = 0, 1, 2, \cdots, N - 1$$

defined over the indicated set of N k-values, where α is a constant and N some positive integer. Suppose we prescribe the *boundary conditions*

$$(4.35) \qquad\qquad y_0 = 0 \qquad y_{N+1} = 0$$

and pose the following *boundary-value problem*:[12] Find all the solutions of (4.34) that simultaneously satisfy the boundary conditions (4.35).

One solution is immediately given by the function which is identically zero for all k-values. This is referred to as a trivial solution and interest centers on the possibility of finding nontrivial solutions of the boundary-value problem. The uniqueness theorem of Section 2.3 does *not* apply here, since y is not prescribed at two *consecutive* k-values. Hence, nontrivial solutions may exist.

The auxiliary equation of (4.34) is

$$(4.36) \qquad\qquad m^2 - (2 - \alpha)m + 1 = 0$$

---

[12] Such boundary-value problems arise in the mathematical theory of scale analysis. For an expository introduction see L. Guttman, "The Principal Components of Scalable Attitudes," Chap. 5 in P. F. Lazarsfeld (ed.), *Mathematical Thinking in the Social Sciences*, Free Press, Glencoe, 1954. The mathematical development may be found in L. Guttman, "The Principal Components of Scale Analysis," Chap. 9 in S. A. Stouffer (ed.), *Measurement and Prediction*, Vol. 4 of *Studies in Social Psychology in World War II*, Princeton Univ. Press, Princeton, 1950.

with roots m_1 and m_2 given, after simplification, by

(4.37)
$$\tfrac{1}{2}[(2 - \alpha) \pm \sqrt{\alpha(\alpha - 4)}].$$

There are three cases, depending upon whether the discriminant of the auxiliary equation is positive (real, unequal roots), zero (real, equal roots), or negative (complex conjugate roots). The discriminant is $\alpha(\alpha - 4)$, which is negative when $0 < \alpha < 4$, zero when $\alpha = 0$ or $\alpha = 4$, and positive for all other values of α.

Suppose $0 < \alpha < 4$. Then roots m_1 and m_2 are complex conjugates, say

(4.38)
$$r(\cos \theta \pm i \sin \theta).$$

The value of r may be found most easily by noting that the product of these roots is just r^2:

$$m_1 m_2 = r(\cos \theta + i \sin \theta)r(\cos \theta - i \sin \theta)$$
$$= r^2(\cos^2 \theta + \sin^2 \theta) = r^2.$$

But the product of the roots of the quadratic equation (4.36) is (footnote, p. 136) equal to 1. Hence $r^2 = 1$ and $r = 1$.

It follows from Theorem 3.7 that the solution y may be written in the form

(4.39)
$$y_k = A \cos (k\theta + B),$$

where A and B are constants. The constants must be evaluated so as to make this solution satisfy the boundary conditions (4.35).

The condition $y_0 = 0$ requires

$$y_0 = A \cos B = 0.$$

Since we seek nontrivial solutions we do not put $A = 0$, but instead satisfy this equation by choosing $B = \pi/2$. Then

$$y_k = A \cos \left(k\theta + \frac{\pi}{2}\right)$$

or

(4.40)
$$y_k = -A \sin k\theta.$$

The second of the boundary conditions (4.35) requires

$$y_{N+1} = -A \sin (N + 1)\theta = 0.$$

Again we avoid $A = 0$ and instead observe that there are values of θ for which we may satisfy this equation. Since $\sin n\pi = 0$ for every integer n, we must choose θ so that

$$(N + 1)\theta = n\pi \quad \text{or} \quad \theta = \frac{n\pi}{N + 1}.$$

The values $n = 0$ and $n = N + 1$ must be omitted since they lead to trivial solutions. If we define

$$(4.41) \qquad \theta_n = \frac{n\pi}{N + 1} \qquad n = 1, 2, \cdots, N,$$

each of these N values of θ determines a nontrivial solution obtained by putting $\theta = \theta_n$ in (4.40). This solution, with $A = -1$, we call $y^{(n)}$. Then

$$(4.42) \qquad y_k^{(n)} = \sin \frac{n\pi k}{N + 1} \qquad n = 1, 2, \cdots, N.$$

Since A is arbitrary in (4.40), any constant multiple of $y^{(n)}$ is also a solution of the boundary-value problem.

Furthermore, owing to the periodicity of the sine function, no additional nontrivial solutions are obtained by letting $n = N + 1$, $N + 2$, $N + 3$, $\cdots$. For example, $n = N + 1$ yields the trivial solution, as does $n = 0$. When $n = N + 2$,

$$\sin \frac{(N + 2)\pi k}{N + 1} = \sin \left(\frac{\pi k}{N + 1} + \pi k \right) = \pm \sin \frac{\pi k}{N + 1}$$

so we get the same set of solutions with $n = N + 2$ as with $n = 1$.

Each value of θ in (4.41) is the amplitude of the corresponding complex conjugate roots of the auxiliary equation (4.36). Each such pair of roots is determined by a particular value of the constant α. The relation between α and θ may be obtained as follows. Since $r = 1$, the sum of the roots given in (4.38) is $2 \cos \theta$. But (footnote, p. 136) the sum of the roots of (4.36) is $2 - \alpha$. Hence

$$2 - \alpha = 2 \cos \theta \quad \text{or} \quad \alpha = 2(1 - \cos \theta).$$

But recall the following identity from trigonometry:

$$\frac{1 - \cos \theta}{2} = \sin^2 \frac{\theta}{2}.$$

Using this, we may write

$$\alpha = 4 \sin^2 \frac{\theta}{2}$$

or, letting α_n be the value of α which corresponds to $\theta = \theta_n$,

$$(4.43) \qquad \alpha_n = 4 \sin^2 \frac{n\pi}{2(N+1)} \qquad n = 1, 2, \cdots, N.$$

Thus, in solving the boundary-value problem for the difference equation (4.34) with boundary conditions (4.35), we are led to nontrivial solutions if α has one of the N-values in (4.43). To $\alpha = \alpha_n$ there corresponds the solution $y^{(n)}$ given in (4.42).

The values of α for which nontrivial solutions of the boundary-value problem exist are called *characteristic values* (or *eigenvalues*) of the problem. The corresponding nontrivial solutions are called *characteristic functions* (or *eigenfunctions*).

We have proceeded from the assumption $0 < \alpha < 4$. We shall soon prove that if α is outside this range of values, no nontrivial solutions of the boundary-value problem exist. Assuming this for the moment, we may summarize our results in the following theorem.

THEOREM 4.3. *The homogeneous second-order difference equation*

$$y_{k+2} - (2 - \alpha)y_{k+1} + y_k = 0 \qquad k = 0, 1, 2, \cdots, N-1$$

with boundary conditions

$$y_0 = 0 \qquad y_{N+1} = 0$$

determines N characteristic values of α given by

$$\alpha_n = 4 \sin^2 \frac{n\pi}{2(N+1)} \qquad n = 1, 2, \cdots, N.$$

To $\alpha = \alpha_n$ there corresponds the characteristic function $y^{(n)}$ given by

$$y_k^{(n)} = \sin \frac{n\pi k}{N+1} \qquad n = 1, 2, \cdots, N.$$

To prove that these N characteristic values are the only ones for this boundary-value problem we have to show that no additional ones are obtained if α is outside the range $0 < \alpha < 4$.

If $\alpha \neq 0$ or $\alpha \neq 4$, the roots of (4.36) are real and unequal and, by Theorem 3.7, the solution of (4.34) is of the form

$$y_k = C_1 m_1^k + C_2 m_2^k.$$

If the first of the boundary conditions (4.35) is to be satisfied,

$$y_0 = C_1 + C_2 = 0$$

or $C_2 = -C_1$. Hence

$$y_k = C_1(m_1^k - m_2^k).$$

For the second boundary condition, we require

(4.44) $$y_{N+1} = C_1(m_1^{N+1} - m_2^{N+1}) = 0.$$

For the quantity in parenthesis to be 0 it is necessary (but not sufficient) that m_1 and m_2 have equal absolute values. Since $m_1 m_2 = 1$, this can happen for real values of the roots only when $m_1 = m_2 = 1$ or $m_1 = m_2 = -1$. But we have assumed $\alpha \neq 0$ and $\alpha \neq 4$ so that m_1 and m_2 are real and unequal. Hence the quantity in parenthesis cannot be 0 and (4.44) is satisfied only when $C_1 = 0$. Thus the trivial solution is the only solution of the boundary problem if $\alpha < 0$ or $\alpha > 4$.

If $\alpha = 0$ or $\alpha = 4$, then $m_1 = m_2$ and the solution becomes

$$y_k = (C_1 + C_2 k)m_1^{\ k}.$$

A similar argument shows that for the boundary conditions (4.35) to be satisfied, both C_1 and C_2 must be 0, so y is again a trivial solution. This completes the argument leading to Theorem 4.3.

The problem just considered is intended to serve as an elementary example to illustrate some important aspects of the theory of characteristic-value problems for difference equations.[13]

PROBLEMS 4.3

1. Show that the difference equation (4.34) may be written in the form

$$\Delta^2 y_k = \alpha E y_k,$$

where Δ and E are the operators (with a difference interval equal to 1) introduced in Chapter 1. (In general, a function is drastically changed when the operator Δ^2 is applied to it, but a characteristic function $y^{(n)}$ is merely multiplied by the characteristic value α_n and displaced 1 unit.)

2. With notation as in Theorem 4.3, show that if

$$r_k = C_1 y_{k+1}^{(1)} + C_2 y_{k+1}^{(2)} + \cdots + C_n y_{k+1}^{(n)},$$

then

$$y_k = \frac{C_1}{\alpha_1} y_k^{(1)} + \frac{C_2}{\alpha_2} y^{(2)} + \cdots + \frac{C_n}{\alpha_n} y_k^{(n)}$$

is a solution of the difference equation $\Delta^2 y_k = r_k$ and also satisfies the boundary conditions (4.35).

3. Consider the boundary-value problem of the text when $N = 2$. Find the characteristic values α_n and the corresponding characteristic solutions.

---

[13] This theory and its applications to problems in physics and engineering are discussed in F. B. Hildebrand, *Methods of Applied Mathematics*, Prentice-Hall, New York, 1952, pp. 249–275. A corresponding theory for *differential equations* is well developed and has many important applications.

4. Consider the difference equation

$$y_{k+2} - y_{k+1} + \alpha y_k = 0 \qquad k = 0, 1, 2, \cdots, N - 1$$

with the boundary conditions

$$y_0 = 0 \qquad y_{N+1} = 0.$$

Follow the method of the text to find a set of characteristic values and functions for this boundary-value problem.

*4.4 GENERATING FUNCTIONS

Powerful methods of analysis, among them the method of generating functions, have been developed to study the particular functions called sequences. As we have seen, the solution of a difference equation defined over the set of k-values $0, 1, 2, \cdots$ is a sequence. It is not surprising therefore to find generating functions extremely useful in solving and studying the solutions of difference equations.[14]

DEFINITION 4.1. *If $y_0, y_1, y_2, \cdots$ is a sequence of real numbers, the function Y (defined for some interval of real numbers containing zero) whose value at s is given by the series*

$$(4.45) \qquad Y(s) = y_0 + y_1 s + y_2 s^2 + \cdots + y_k s^k + \cdots$$

is the generating function of the sequence $\{y_k\}$.

Examples

(a) The function given by

$$Y(s) = \frac{1}{1 - s}$$

is the generating function of the constant sequence $1, 1, 1, \cdots$. For if we perform the division of 1 by $(1 - s)$, we find

$$(4.46) \qquad \frac{1}{1 - s} = 1 + s + s^2 + \cdots + s^k + \cdots.$$

Hence, in view of (4.45), $y_0 = 1, y_1 = 1, y_2 = 1, \cdots$.

(b) Again by a long division [or, for those familiar with infinite series, by forming the Cauchy product of the series (4.46) with itself, or by termwise differentiation of (4.46)], we calculate

$$(4.47) \qquad \frac{1}{(1 - s)^2} = 1 + 2s + 3s^2 + \cdots + (k + 1)s^k + \cdots.$$

---

[14] The reader may consult almost any calculus textbook for those elementary aspects of the theory of infinite series used in this section. That same textbook will also include the rules for differentiating power and composite functions needed in some of the illustrative material from the theory of probability.

Hence the generating function of the sequence $\{k + 1\}$ or $1, 2, 3, \cdots$ is given by $1/(1 - s)^2$. By multiplying (4.47) by s we obtain the series $s + 2s^2 + 3s^3 + \cdots$ and thus are able to identify $s/(1 - s)^2$ as the generating function of the sequence $\{k\}$ or $0, 1, 2, \cdots$.

(c) To find the generating function of the sequence $\{A^k\}$ where $A \neq 0$ is a constant, we observe that

$$Y(s) = 1 + As + A^2s^2 + \cdots + A^ks^k + \cdots$$

is an infinite geometric series with a common ratio equal to As. If s is restricted to the interval $|s| < 1/|A|$, then this ratio is less than 1 in absolute value and we may apply the formula[15] for the sum of an infinite geometric progression to obtain

$$Y(s) = \frac{1}{1 - As}.$$

(d) From the binomial theorem [see (1.42)] we find

$$(A + Bs)^n = \sum_{k=0}^{n} \binom{n}{k} A^{n-k}B^ks^k.$$

Hence, for any positive integer n, $(A + Bs)^n$ is the generating function of the sequence $\{y_k\}$ for which $y_k = \binom{n}{k} A^{n-k}B^k$ if $k = 0, 1, 2, \cdots, n$ and $y_k = 0$ if $k > n$.

Table 4.1 summarizes these calculations. From this table we can find the generating function of any of the special sequences on the left as well as identify the sequence if one of the generating functions on the right is given.

TABLE 4.1

Line	General Term of Sequence $\{y_k\}$	Generating Function $Y(s)$
(1)	1	$\dfrac{1}{1 - s}$
(2)	k	$\dfrac{s}{(1 - s)^2}$
(3)	$k(k + 1)$	$\dfrac{2s}{(1 - s)^3}$
(4)	A^k	$\dfrac{1}{1 - As}$
(5)	$\begin{cases} \binom{n}{k} A^{n-k}B^k & k = 0, 1, \cdots, n \\ 0 & k > n \end{cases}$	$(A + Bs)^n$

[15] If $-1 < r < 1$, the infinite geometric series $a + ar + ar^2 + \cdots$ has sum $a/(1 - r)$. This is the limit as $n \to \infty$ of the sum of the first n terms of this series. [See the formula for this partial sum in 15 of Problems 1.6.]

It is useful to think of the calculation of a generating function as a *transformation* from the sequence (i.e., a function whose domain of definition is the set of integers 0, 1, 2, · · ·) to its generating function (i.e., a function whose domain is a set of real numbers in some interval containing $s = 0$). We read the result of this transformation by moving from left to right on any line of the above table. Moving in the opposite direction, i.e., from a given generating function to its corresponding sequence, is known as performing the *inverse* transformation.

The fundamental property of the generating function transformation is its *linearity: if $\{y_k^{(1)}\}$ has the generating function Y_1 and $\{y_k^{(2)}\}$ has the generating function Y_2, then $\{c_1 y_k^{(1)} + c_2 y_k^{(2)}\}$ has the generating function $c_1 Y_1 + c_2 Y_2$, for any constants c_1 and c_2.* For a proof of this property, we need only add the two given equations

$$Y_1(s) = y_0^{(1)} + y_1^{(1)}s + y_2^{(1)}s^2 + \cdots + y_k^{(1)}s^k + \cdots$$
$$Y_2(s) = y_0^{(2)} + y_1^{(2)}s + y_2^{(2)}s^2 + \cdots + y_k^{(2)}s^k + \cdots$$

after multiplying the first by c_1 and the second by c_2. (This termwise multiplication and addition of series, although not generally valid operations, are permissible for power series.) We obtain

$$
\begin{aligned}
c_1 Y_1(s) + c_2 Y_2(s) = (c_1 y_0^{(1)} + c_2 y_0^{(2)}) &+ (c_1 y_1^{(1)} + c_2 y_1^{(2)})s \\
&+ (c_1 y_2^{(1)} + c_2 y_2^{(2)})s^2 + \cdots \\
&+ (c_1 y_k^{(1)} + c_2 y_k^{(2)})s^k + \cdots,
\end{aligned}
$$

which identifies $c_1 Y_1 + c_2 Y_2$ as the generating function of the sequence with general term $c_1 y_k^{(1)} + c_2 y_k^{(2)}$.

We note one other result before applying generating functions to the study of difference equations. If the generating function of the sequence $\{y_k\}$ is known, we may easily find the generating functions of the sequences $\{y_{k+1}\}$, $\{y_{k+2}\}$, · · ·. For starting with (4.45) we find

$$(4.48) \qquad \frac{Y(s) - y_0}{s} = y_1 + y_2 s + y_3 s^2 + \cdots + y_{k+1} s^k + \cdots$$

so that the sequence $\{y_{k+1}\}$ has a generating function given by the left-hand member of (4.48). Similarly, one finds that the generating function of $\{y_{k+2}\}$ is given by

$$\frac{Y(s) - y_0 - y_1 s}{s^2} = y_2 + y_3 s + y_4 s^2 + \cdots + y_{k+2} s^k + \cdots.$$

These results, given in Table 4.2, are a continuation of the five entries in Table 4.1.

The use of generating functions in solving difference equations arises from two facts: (1) often the generating function of the solution sequence is more easily obtained than the solution itself, and (2) with its generating function known, the solution sequence can be explicitly calculated (by performing the inverse transformation mentioned above); even if this

TABLE 4.2

Line	General Term of Sequence	Generating Function
(6)	y_k	$Y(s)$
(7)	y_{k+1}	$\dfrac{Y(s) - y_0}{s}$
(8)	y_{k+2}	$\dfrac{Y(s) - y_0 - y_1 s}{s^2}$
(9)	y_{k+n}	$\dfrac{Y(s) - y_0 - y_1 s - \cdots - y_{n-1} s^{n-1}}{s^n}$
(10)	$c_1 y_k^{(1)} + c_2 y_k^{(2)}$	$c_1 Y_1(s) + c_2 Y_2(s)$

calculation is not feasible, an analysis of its generating function may indirectly reveal important properties of the solution.

In order to illustrate these remarks without excessive computational difficulties we shall apply the generating function technique to the first-order difference equation

$$(4.49) \qquad y_{k+1} = A y_k + B \qquad k = 0, 1, 2, \cdots,$$

where A and B are constants. With y_0 prescribed, we know (Theorem 2.7) the unique solution of (4.49) to be

$$(4.50) \qquad y_k = \begin{cases} A^k(y_0 - y^*) + y^* & \text{if } A \neq 1 \\ y_0 + Bk & \text{if } A = 1 \end{cases}$$

where

$$(4.51) \qquad y^* = \frac{B}{1 - A}.$$

We want now to obtain (4.50) via the method of generating functions. Letting Y denote the generating function of $\{y_k\}$, we read from line (7) in Table 4.2 the generating function of $\{y_{k+1}\}$ and from lines (10) and (1) of Tables 4.1 and 4.2 we find $A Y(s) + B [1/(1 - s)]$ as the generating function of $\{A y_k + B\}$. Hence, (4.49) is transformed into

$$\frac{Y(s) - y_0}{s} = A Y(s) + \frac{B}{1 - s}$$

which when simplified becomes

$$(4.52) \qquad Y(s) = \frac{y_0}{1 - As} + \frac{Bs}{(1 - s)(1 - As)}.$$

This is the generating function of the solution sequence of the difference equation (4.49) and our task is to perform the inverse transformation, i.e., identify the sequence $\{y_k\}$ having (4.52) as its generating function. From line (4) of Table 4.1, we immediately recognize $\{y_0 A^k\}$ as the sequence having $y_0/(1 - As)$ as its generating function. However, the second term on the right side of (4.52) must first be simplified somewhat before we are able to identify its corresponding sequence. To do this, we use the method of *partial fractions*,[16] i.e., we determine constants α and β for which

$$(4.53) \qquad \frac{s}{(1 - s)(1 - As)} = \frac{\alpha}{1 - s} + \frac{\beta}{1 - As}$$

is an identity. To find α and β we write

$$\frac{s}{(1 - s)(1 - As)} = \frac{\alpha(1 - As) + \beta(1 - s)}{(1 - s)(1 - As)}$$

and then equate numerators to obtain

$$s = (\alpha + \beta) + (-\alpha A - \beta)s.$$

For this to be true for all values of s, α and β must satisfy the equations

$$\alpha + \beta = 0 \qquad -\alpha A - \beta = 1.$$

From the first equation $\beta = -\alpha$, and the second equation becomes $\alpha(1 - A) = 1$. Hence, if $A \neq 1$,

$$\alpha = \frac{1}{1 - A} \qquad \beta = -\frac{1}{1 - A},$$

and recalling (4.53)

$$\frac{s}{(1 - s)(1 - As)} = \frac{1}{1 - A}\left(\frac{1}{1 - s} - \frac{1}{1 - As}\right) \qquad A \neq 1.$$

We are now able to write the generating function (4.52) in the form

$$Y(s) = \frac{y_0}{1 - As} + \frac{B}{1 - A}\left(\frac{1}{1 - s} - \frac{1}{1 - As}\right) \qquad A \neq 1,$$

---

[16] The technique of decomposing a rational function into partial fractions is outlined in most calculus textbooks. A proof of the validity of this technique is found in C. C. MacDuffee, *Theory of Equations*, John Wiley, New York, 1954, pp. 41–44.

or using (4.51) and collecting terms, in the still more convenient form

$$Y(s) = (y_0 - y^*)\frac{1}{1 - As} + y^*\frac{1}{1 - s}.$$

Now, moving from right to left on lines (10), (4), and (1) of our tables of generating functions, we obtain

$$y_k = (y_0 - y^*)A^k + y^* \qquad A \neq 1.$$

This is the solution (4.50) in the case $A \neq 1$. To complete the argument, note that if $A = 1$ the generating function (4.52) becomes

$$Y(s) = \frac{y_0}{1 - s} + \frac{Bs}{(1 - s)^2}$$

which, directly from the first two lines of Table 4.1, leads to the solution

$$y_k = y_0 + Bk,$$

again as in (4.50).

The following example applies the method of generating functions to solve a second-order difference equation.

Example

If we are given

$$y_{k+2} - 2y_{k+1} + y_k = 2^k$$

with $y_0 = 2$ and $y_1 = 1$ prescribed, then directly from lines (8), (7), (6), and (4) of our tables of generating functions we find

$$\frac{Y(s) - 2 - s}{s^2} - 2\frac{Y(s) - 2}{s} + Y(s) = \frac{1}{1 - 2s}.$$

To simplify, we multiply through by s^2, collect terms, and find

$$(1 - s)^2 Y(s) = 2 - 3s + \frac{s^2}{1 - 2s}$$

or

$$Y(s) = \frac{2}{(1 - s)^2} - \frac{3s}{(1 - s)^2} + \frac{s^2}{(1 - s)^2(1 - 2s)}.$$

We must reduce this last term by partial fractions:

$$\frac{s^2}{(1 - s)^2(1 - 2s)} = \frac{\alpha}{1 - s} + \frac{\beta}{(1 - s)^2} + \frac{\gamma}{1 - 2s}$$

$$= \frac{\alpha(1 - s)(1 - 2s) + \beta(1 - 2s) + \gamma(1 - s)^2}{(1 - s)^2(1 - 2s)}$$

$$= \frac{(\alpha + \beta + \gamma) + (-3\alpha - 2\beta - 2\gamma)s + (2\alpha + \gamma)s^2}{(1 - s)^2(1 - 2s)}$$

Since these numerators must be identical, we obtain the three simultaneous linear equations for α, β, and γ:

$$\alpha + \beta + \gamma = 0, \qquad -3\alpha - 2\beta - 2\gamma = 0, \qquad 2\alpha + \gamma = 1,$$

from which $\alpha = 0$, $\beta = -1$, and $\gamma = 1$. Hence

$$Y(s) = \frac{1}{(1-s)^2} - \frac{3s}{(1-s)^2} + \frac{1}{1-2s}$$

and from our tables of generating functions, we identify the solution

$$y_k = (k+1) - 3k + 2^k = 1 - 2k + 2^k.$$

In general, the method of generating functions is outlined as follows: (1) A difference equation is given together with the prescribed initial condition(s), and its solution, a sequence $\{y_k\}$, is to be found. (2) The generating function transformation is applied to the difference equation which thereby is transformed into an algebraic equation easily solved (with the aid of the initial conditions) for Y, the generating function of the solution sequence $\{y_k\}$. (3) The inverse transformation is then performed on Y in order to obtain the desired solution $\{y_k\}$. Much of the theory of difference equations, in particular equations with constant coefficients, can be developed using such generating function techniques. Of course, this requires a much more extensive knowledge of pairs of sequences and their generating functions than that summarized in the ten lines of Tables 4.1 and 4.2.[17]

Other transformations, e.g., those named after Laplace, Dirichlet, and Laurent, have also been used in this way to solve and study not only difference equations[18] but also differential, integral, and other functional equations. We shall not pursue these matters here but instead return to the method of generating functions and some questions in the theory of probability where this method has been found extremely useful.[19]

[17] See C. Jordan, *Calculus of Finite Differences*, Budapest, 1939, for the calculation of many generating functions (pp. 20–44) and for the application of the generating function method to the solution of linear difference equations (pp. 572–576, 586–587).

[18] For applications of the Laplace transform to difference equations see R. V. Churchill, *Modern Operational Mathematics in Engineering*, McGraw-Hill, New York, 1944, pp. 23–28; M. F. Gardner and J. L. Barnes, *Transients in Linear Systems*, Vol. 1, John Wiley, New York, 1942, Chap. IX; E. J. Scott, *Transform Calculus*, Harper, New York, 1955, Chap. VI. For the Dirichlet transform see T. Fort, "Linear Difference Equations and the Dirichlet Series Transform," *American Mathematical Monthly*, *62* (1955), 641–645. For the Laurent transform see D. F. Lawden, "On the Solution of Linear Difference Equations," *Mathematical Gazette*, *36* (1952), 193–196.

[19] In fact, generating functions were first introduced by Laplace in his *Théorie analytique des probabilités*, Paris, 1812. They are used to solve some special problems in J. V. Uspensky, *Introduction to Mathematical Probability*, McGraw-Hill, New York, 1937, and are systematically developed and applied in W. Feller, *An Introduction to Probability Theory and Its Applications*, Vol. 1, 2nd ed., John Wiley, New York, 1957, especially Chap. XI.

Consider an experiment each of whose outcomes has been unambiguously classified as either a success (S) or a failure (F). For example, in tossing a coin we may call heads a success and tails a failure; in running a subject through a T-maze, we may label a turn to the right an S, to the left an F; in asking a question in a poll, we may call the answer "yes" an S, any other answer ("no," "don't know," $\cdots$) an F. Let the probability of a success be denoted by p. Then $0 \leq p \leq 1$ and the probability of a failure is given by $q = 1 - p$.

Now suppose this experiment is repeated a certain number of times, say n, each trial being independent of the others. Since each experiment results in an S or an F, repeating the experiment produces a sequence of S's and F's. Thus in three tosses of a coin, the result heads, tails, heads, in that order, may be denoted by SFS. We want to study the variable X defined as the *total number of successes* obtained in the n repetitions of the experiment.

X is called a *random variable*: it may assume any one of the values $0, 1, 2, \cdots, n$ each with a certain probability which measures the relative frequency with which the n repetitions lead to 0 or 1 or 2, $\cdots$, or n successes. We denote by $y_{k,n}$ the probability that the number of successes in the n repetitions will be exactly k, i.e.,

(4.54) $$y_{k,n} = P(X = k) \qquad k = 0, 1, 2, \cdots, n,$$

where we write $P(\ )$ as shorthand for the probability of the event specified in parentheses. Of course, we are certain (hence the probability is 1) that we will obtain exactly one of the numbers $0, 1, 2, \cdots, n$ as the value of X each time we do these n experiments, i.e.,

(4.55) $$\sum_{k=0}^{n} y_{k,n} = y_{0,n} + y_{1,n} + y_{2,n} + \cdots + y_{n,n} = 1.$$

The $(n + 1)$ numbers $y_{0,n}, y_{1,n}, \cdots, y_{n,n}$ defined by (4.54) tell us the way in which this total probability of 1 is distributed among the $(n + 1)$ possible different values of X. For this reason, the function y is called the *probability distribution* of X. The parameter n being fixed, y is defined for the set of k-values $0, 1, 2, \cdots, n$. To find y we derive a difference equation which it satisfies and then solve this equation using generating functions.

The event that $(k + 1)$ successes occur in $(n + 1)$ repetitions of the experiment is possible only in the following two mutually exclusive ways: (1) by having $(k + 1)$ successes after n repetitions and then a failure in the $(n + 1)$st experiment, or (2) by having k successes after n repetitions and then a success in the $(n + 1)$st experiment. The probability of the event in (1) is the *product* of the probability of $(k + 1)$ successes in the first n experiments and the probability of a failure in the $(n + 1)$st. (This follows from the multiplication rule for the probability of the joint occurrence of two *independent* events.) The probability of the event in (1) is therefore $qy_{k+1,n}$. Similarly, the probability of the event in (2) is $py_{k,n}$. Hence (by the addition rule for the probability of the occurrence of one or the other of two *mutually exclusive* events)

$$(4.56) \quad y_{k+1,n+1} = qy_{k+1,n} + py_{k,n}$$
$$k = 0, 1, 2, \cdots, n \quad n = 0, 1, 2, \cdots.$$

The above argument breaks down when $k = -1$ since the event that zero successes occur in $(n + 1)$ repetitions is possible only by having zero successes after n repetitions and then a failure in the $(n + 1)$st experiment. Hence we obtain the equation

$$(4.57) \quad y_{0,n+1} = qy_{0,n} \quad n = 0, 1, 2, \cdots.$$

Finally, we note the initial conditions

$$(4.58) \quad y_{0,0} = 1 \quad y_{k,0} = 0 \quad \text{if } k > 0$$

which express the fact that in 0 repetitions of the experiment it is impossible to have any number of successes other than zero.

Equation (4.56) is a *partial difference equation* (see 13 and 14 of Problems 4.4) for the function y. Together with the initial conditions (4.57) and (4.58), this difference equation can be solved to find y, the probability distribution of the random variable X.

If we consider n an arbitrary but fixed integer, then $y_{k,n}$ may be thought of as the general term of the sequence $\{y_{k,n}\}$ and we may therefore define the generating function Y_n of this sequence as follows:

$$(4.59) \quad Y_n(s) = y_{0,n} + y_{1,n}s + y_{2,n}s^2 + \cdots + y_{k,n}s^k + \cdots.$$

There is one such generating function for each value of n. Now we apply the generating function transformation to the difference equation (4.56). We use line (7) of Table 4.2 to obtain

$$\frac{Y_{n+1}(s) - y_{0,n+1}}{s} = q\frac{Y_n(s) - y_{0,n}}{s} + pY_n(s) \quad n = 0, 1, 2, \cdots$$

which when simplified (by multiplying through by s and collecting terms) becomes

$$Y_{n+1}(s) = (q + ps)Y_n(s) + (y_{0,n+1} - qy_{0,n}) \quad n = 0, 1, 2, \cdots.$$

Because of condition (4.57), this may be further simplified to

(4.59') $Y_{n+1}(s) = (q + ps) Y_n(s)$ $n = 0, 1, 2, \cdots$

which is a first-order difference equation for the generating function Y. From (4.58) we note that

(4.60) $Y_0(s) = y_{0.0} + y_{1.0}s + y_{2.0}s^2 + \cdots = y_{0.0} = 1.$

The solution of (4.59') subject to the initial condition (4.60) can be obtained from Theorem 2.7 (with $A = q + ps$, $B = 0$). We find

(4.61) $Y_n(s) = (q + ps)^n$ $n = 0, 1, 2, \cdots.$

To calculate the probability distribution function given by the sequence $\{y_{k,n}\}$ we need only apply the inverse transformation to the generating function in (4.61). Directly from line (5) of Table 4.1, we obtain

(4.62) $y_{k,n} = P(X = k) = \binom{n}{k} p^k q^{n-k}$ $k = 0, 1, 2, \cdots, n$

but $y_{k,n} = 0$ if $k > n$.

The probabilities specified in (4.62) define the so-called *binomial* probability distribution (with parameters n and p) and a random variable having this probability distribution is said to be binomially distributed. Thus we may summarize our calculations as follows: *the number of successes in n independent repetitions of an experiment allowing only the two outcomes success (with probability p) and failure (with probability $q = 1 - p$) is binomially distributed (with parameters n and p), i.e., the probability that this number of successes is exactly k is given by* (4.62).

Examples

(a) If, in tossing a fair die, we regard a 1 or 2 as success and any other number as failure, then the probability of success in any one toss is $\frac{1}{3}$ and the probability of exactly two successes in four independent tosses of the die is

$$y_{2.4} = P(X = 2) = \binom{4}{2} \left(\frac{1}{3}\right)^2 \left(\frac{2}{3}\right)^2 = \frac{4!}{2!2!} \cdot \frac{1}{9} \cdot \frac{4}{9} = \frac{8}{27}.$$

(b) In running ten rats through a maze, it is observed that eight turn right and only two turn left at the first choice-point. If it is hypothesized that each rat is equally likely to turn right or left and does so independently of the others, what is the probability P of obtaining at least this many turning right? We have an experiment whose outcomes are dichotomized into success (right turn) and failure (left turn). This experiment is repeated ten independent times with the constant probability of success equal to $\frac{1}{2}$. The number of right turns is therefore binomially distributed (with parameters $n = 10$ and $p = \frac{1}{2}$). **The** probability of at least eight successes is the probability of eight, nine, or ten successes.   Hence

$$P = y_{8,10} + y_{9,10} + y_{10,10}$$

$$= \binom{10}{8} \left(\frac{1}{2}\right)^8 \left(\frac{1}{2}\right)^2 + \binom{10}{9} \left(\frac{1}{2}\right)^9 \left(\frac{1}{2}\right)^1 + \binom{10}{10} \left(\frac{1}{2}\right)^{10}$$

$$= \frac{45}{1024} + \frac{10}{1024} + \frac{1}{1024} = \frac{56}{1024} = 0.055, \text{ approximately.}$$

Calculations of this kind are basic in many statistical problems and extensive tables of the binomial distribution probabilities (4.62) are available.[20]   Many properties of the binomial distribution may be obtained directly from the generating function (4.61).   We note here only one result, leaving some others for the problems.

The *mean value*, $E(X)$, of a random variable X, sometimes referred to as the average value, expected value, or mathematical expectation of X, is defined as follows: if X assumes the values $x_0, x_1, x_2, \cdots$ with probabilities $p_0, p_1, p_2, \cdots$ respectively, then

$$(4.63) \qquad\qquad E(X) = \sum_{k=0} x_k p_k,$$

where the summation is extended over all values of X.   (This definition applies only to discrete random variables, i.e., those whose values form a finite or infinite sequence.)   In words, $E(X)$ is the weighted average cf the values of X, each weighted by the probability of its occurrence.   For example, if a random variable can assume only the two values 1 and 3. each with probability $\frac{1}{2}$, then $E(X) = 1 \cdot \frac{1}{2} + 3 \cdot \frac{1}{2} = 2$.

We want to use the generating function (4.61) to prove that *if X is binomially distributed (with parameters n and p), then $E(X) = np$*.   From the definition of mean value, since X assumes the value k with probability $y_{k,n}$ and k goes from 0 to n,

$$(4.64) \qquad\qquad E(X) = \sum_{k=0}^{n} k y_{k,n}.$$

We must prove this sum equal to np.   Directly from the definition of the generating function $Y_n(s)$ in (4.59), differentiation with respect to s leads to (with a prime indicating differentiation)

$$Y_n'(s) = y_{1,n} + 2y_{2,n}s + \cdots + k y_{k,n} s^{k-1} + \cdots$$

which when s is put equal to 1 is seen to reduce to precisely the sum (4.64).

---

[20] For example, *Tables of the Cumulative Binomial Probability Distribution*, Harvard Univ. Press, Cambridge, 1955.   Section IV of the Introduction, written by F. Mosteller, illustrates 17 different statistical applications of the binomial distribution.

Hence to find the mean value of X we need only find $Y_n'(1)$, the value of the derivative of the generating function Y_n when $s = 1$. From (4.61) we easily calculate $Y_n'(s) = n(q + ps)^{n-1}p$ and (remembering that $p + q = 1$)

$$E(X) = n(q + p)^{n-1}p = np,$$

as claimed.

Examples

(a) The mean number of rats turning right at a choice-point in a maze when ten rats, each equally likely to turn right or left, are independently run through the maze is $10 \cdot \frac{1}{2}$ or 5.

(b) If 1000 people are sampled and it is assumed that each answers independently of the others and each person answers "yes" with probability $\frac{1}{4}$ and "no" with probability $\frac{3}{4}$, then the mean or expected number of people who answer "yes" in such samples of size 1000 is $1000 \cdot \frac{1}{4}$ or 250.

We conclude this section with an application of generating functions to a mathematical model of verbal learning.[21] The experimental procedure is as follows: A subject is presented with a list of words after which he writes down all the words he can remember. This procedure is then repeated on successive trials, the order of the words being randomized before each presentation.

A word will be said to be in state k if it has been recalled exactly k times on the preceding trials. Let $p_{k,n} =$ probability that a word is in state k on trial number n and $\tau_k =$ probability that a word in state k is recalled, with both k and n assuming the values $0, 1, 2, \cdots$. The fact that τ_k is defined to be independent of the trial number n means that the following assumption has been made: "The degree to which any word in the test material has been learned is completely specified by the number of times the word has been recalled on the preceding trials." (Miller and McGill, p. 369.) The numbers $\tau_0, \tau_1, \tau_2, \cdots$ are estimated from the experimental data and we shall assume them known. The probabilities $p_{k,n}$ are to be determined. The first step is the derivation of a partial difference equation satisfied by these probabilities.

The event that a word is in state $k + 1$ on trial number $n + 1$ can occur in only two mutually exclusive ways: either (1) the word is in state $k + 1$ on trial n and is not recalled on trial $n + 1$, or (2) the word is in state k on trial n and is recalled on trial $n + 1$. If we assume successive trials are statistically independent, then the probability of event (1) is the product of $p_{k+1,n}$, the probability that the word is in state $k + 1$ on trial n, and $(1 - \tau_{k+1})$, the probability that this word is not recalled on the following trial. Similarly, event (2) has as its probability the product of $p_{k,n}$ and τ_k. Hence

$$(4.65) \qquad p_{k+1,n+1} = (1 - \tau_{k+1})p_{k+1,n} + \tau_k p_{k,n}$$
$$k = 0, 1, 2, \cdots; \quad n = 0, 1, 2, \cdots.$$

Note that the above argument does not apply to $p_{0,n+1}$ since a word can be in state 0 (i.e., never recalled) on trial $n + 1$ only if it is in state 0 on trial n and then is not recalled on trial $n + 1$. Hence

$$(4.66) \qquad p_{0,n+1} = (1 - \tau_0)p_{0,n} \qquad n = 0, 1, 2, \cdots.$$

A set of initial conditions may be written expressing the fact that no words are recalled before the first trial, i.e.,

$$(4.67) \qquad p_{0,0} = 1 \qquad p_{k,0} = 0 \qquad \text{if } k = 1, 2, 3, \cdots.$$

Equation (4.65) is a partial difference equation somewhat more difficult to solve, but of the same general form as (4.56). Miller and McGill use matrix analysis to solve (4.65) but we shall apply the method of generating functions.[22]

If k is considered fixed and $p_{k,n}$ viewed as the general term of the sequence $p_{k,0}, p_{k,1}, p_{k,2}, \cdots$, we can define the generating function

$$(4.68) \qquad P_k(s) = \sum_{n=0}^{\infty} p_{k,n} s^n.$$

This generating function transformation is now applied to both (4.65) and (4.66). Taking (4.66) first [and using line (7) of Table 4.2] we obtain

$$\frac{P_0(s) - p_{0,0}}{s} = (1 - \tau_0)P_0(s)$$

which, in view of the first initial condition in (4.67), simplifies to

$$(4.69) \qquad P_0(s) = \frac{1}{1 - s(1 - \tau_0)}.$$

In a similar manner we transform (4.65) into

$$\frac{P_{k+1}(s) - p_{k+1,0}}{s} = (1 - \tau_{k+1})P_{k+1}(s) + \tau_k P_k(s) \qquad k = 0, 1, 2, \cdots$$

---

[22] This is the method used in M. A. Woodbury, "On a Probability Distribution," *Annals of Mathematical Statistics*, 20 (1949), 311–313, to which the reader is referred for results or proofs omitted here.

which, again using (4.67), becomes

$$(4.70) \qquad P_{k+1}(s) = \frac{s\tau_k}{1 - s(1 - \tau_{k+1})} P_k(s) \qquad k = 0, 1, 2, \cdots.$$

Equation (4.70) is a homogeneous, first-order difference equation which can be solved by the method used in Section 4.2 to solve (4.22). We find

$$(4.71) \qquad P_k(s) = P_0(s) \prod_{j=0}^{k-1} \frac{s\tau_j}{1 - s(1 - \tau_{j+1})} \qquad k = 1, 2, 3, \cdots$$

and this, together with (4.69) for $P_0(s)$, determines (for each value of k) the generating function P_k of the sequence $\{p_{k,n}\}$.

The problem of finding the probabilities $p_{k,n}$ is now reduced to that of performing the inverse transformation on the generating functions in (4.69) and (4.71). This is easy to do in the case $k = 0$ for, directly from line (4) of Table 4.1, we invert (4.69) to obtain

$$(4.72) \qquad p_{0,n} = (1 - \tau_0)^n \qquad n = 0, 1, 2, \cdots.$$

Putting $k = 1$ in (4.71) we find

$$(4.73) \qquad P_1(s) = \frac{\tau_0 s}{[1 - s(1 - \tau_0)][1 - s(1 - \tau_1)]}$$

which, by using the method of partial fractions, may be written in the equivalent form

$$(4.74) \; P_1(s) = \frac{\tau_0}{\tau_1 - \tau_0} \left\{ \frac{1}{1 - s(1 - \tau_0)} - \frac{1}{1 - s(1 - \tau_1)} \right\} \qquad \tau_1 \neq \tau_0.$$

Now we use lines (4) and (10) of the tables of generating functions to invert and thus to obtain, if $\tau_1 \neq \tau_0$,

$$(4.75) \qquad p_{1,n} = \frac{\tau_0}{\tau_1 - \tau_0} [(1 - \tau_0)^n - (1 - \tau_1)^n].$$

The probabilities $p_{k,n}$ for $k = 2, 3, 4, \cdots$ can also be explicitly found, but we do not reproduce these here. Very often information may be directly obtained from the generating function, thus obviating the necessity of finding explicit solutions.

To illustrate this point, suppose we ask for the probability that a word is recalled k times and then is never recalled again as repeated trials take place. The product $p_{k,n}\tau_k$ is the probability that a word is recalled k times preceding trial number n and then is recalled. Since n may assume any integer value, if we form the sum

$$\pi_k = \sum_{n=0}^{\infty} p_{k,n}\tau_k$$

we are finding the probability that a word is recalled k times and then *is* recalled again. Since τ_k is independent of the summation variable n, we may write

$$(4.76) \qquad \pi_k = \tau_k \sum_{n=0}^{\infty} p_{k,n}.$$

But the sum in (4.76) is precisely the value when $s = 1$ of the generating function P_k defined in (4.68). Hence

$$(4.77) \qquad \pi_k = \tau_k P_k(1)$$

and we need only calculate $P_k(1)$ to evaluate the probability π_k. From (4.69) we find $P_0(1) = 1/\tau_0$ (assuming $\tau_0 \neq 0$) so that

$$\pi_0 = \tau_0 P_0(1) = 1.$$

From (4.71) we likewise calculate (assuming $\tau_0 \neq 0,\ \tau_1 \neq 0,\ \cdots,$ $\tau_k \neq 0$),

$$P_k(1) = P_0(1) \prod_{j=0}^{k-1} \frac{\tau_j}{1 - (1 - \tau_{j+1})}$$

$$= \frac{1}{\tau_0} \prod_{j=0}^{k-1} \frac{\tau_j}{\tau_{j+1}}$$

$$= \frac{1}{\tau_0} \cdot \frac{\tau_0}{\tau_1} \cdot \frac{\tau_1}{\tau_2} \cdot \frac{\tau_2}{\tau_3} \cdots \frac{\tau_{k-2}}{\tau_{k-1}} \cdot \frac{\tau_{k-1}}{\tau_k}$$

$$= \frac{1}{\tau_k},$$

since the denominator of each fraction but the last is canceled by the numerator of the following fraction. Hence, recalling (4.77), if all the transition probabilities τ_k ($k = 0, 1, 2, \cdots$) are positive, then

$$(4.78) \qquad \pi_k = 1 \qquad k = 0, 1, 2, \cdots,$$

i.e., a word in state k has probability 1 to move to state $k + 1$ if the learning trials are indefinitely continued. It follows that in the limit (as the number of trials increases without bound) the probability of any *finite* number of recalls is 0.

For further aspects of this learning model the reader is referred to the original paper cited in footnote 21. Additional material on free-recall verbal learning may also be found in Bush and Mosteller's book.[23]

---

[23] R. R. Bush and F. Mosteller, *Stochastic Models for Learning*, John Wiley, New York, 1955, Chap. 10.

PROBLEMS 4.4

1. Find the generating function of the sequence $\{y_k\}$ if

(a) $y_k = 2 + 3k$

(d) $y_k = (k + 1)(k + 2)$

(b) $y_k = p^k q$

(e) $y_k = \dfrac{k(k - 1)}{2}$

(c) $y_0 = y_1 = 0, y_k = 1$ if $k > 1$

(f) $y_0 = 0, y_1 = 1, y_k = 0$ if $k > 1$.

2. If $\{y_k\}$ has a generating function $Y(s)$, show that the generating function of $\{\Delta y_k\}$ where $\Delta y_k = y_{k+1} - y_k$ is given by $(1/s)[(1 - s)Y(s) - y_0]$.

3. Find the sequence $\{y_k\}$ having the generating function Y given by

(a) $Y(s) = \dfrac{3}{1 - s} + \dfrac{1}{1 - 2s}$

(d) $Y(s) = \dfrac{s}{1 - 2s}$

(b) $Y(s) = \dfrac{s}{(1 - s)^2} + \dfrac{s}{1 - s}$

(e) $Y(s) = (1 + s)^4$

(c) $Y(s) = \dfrac{s^2}{1 - s}$

(f) $Y(s) = (3 - s)^n$

4. Consider the difference equation

$$y_{k+1} = y_k + k \qquad k = 0, 1, 2, \cdots$$

with the initial condition $y_0 = 1$. (a) By applying the generating function transformation to this difference equation, show that the generating function of the solution sequence $\{y_k\}$ is given by

$$Y(s) = \frac{s^2}{(1 - s)^3} + \frac{1}{1 - s}.$$

(b) Apply the inverse transformation to find the solution

$$y_k = \frac{k(k - 1)}{2} + 1.$$

5. Use the method of generating functions to find the solution of the difference equation

$$y_{k+2} - 5y_{k+1} + 6y_k = 2$$

with initial conditions $y_0 = 1, y_1 = 2$.

6. The deck of cards used in extrasensory perception (ESP) experiments consists of 25 cards—5 cards of each of 5 different designs. A guess is made as to the design of a single card selected at random from this deck. This card is then replaced, the deck shuffled, and another guess made as to the design of another card selected. This procedure is repeated for a total of five guesses. Use the binomial probabilities (4.62) to calculate the probability of (a) exactly four correct guesses, (b) at least four correct guesses. Show that the mean value of the number of correct guesses is 1.

7. Use the generating function found in (4.61) to prove property (4.55) of the sequence $\{y_{k,n}\}$.

8. A random variable X has a *geometric* probability distribution if it can assume any positive integer value with probabilities given by

$$P(X = k) = p^{k-1}q \qquad k = 1, 2, 3, \cdots,$$

p and q being positive constants with $p = 1 - q$. Show that the generating function of this distribution is $Y(s) = qs/(1 - ps)$ and then show by computing $Y'(1)$ that $E(X) = 1/q$. [$P(X = k)$ is the probability that the *first* failure in a sequence of independent trials (of the type described on p. 196) occurs at the kth trial. We tacitly agree to put $P(X = 0) = 0$.]

9. In independent repetitions of an experiment allowing only the two outcomes success (with probability p) and failure (with probability $q = 1 - p$), let X denote the length of a run[24] of successes, i.e., the number of consecutive successes initiated by a specified success. For example, in the sequence *SSFSSSFSF* there are three success runs of lengths 2, 3, and 1 and initiated by successes at trial numbers 1, 4, and 8 respectively. (*a*) Show that the random variable X has the geometric distribution defined in Problem 8. (*b*) What is the mean or expected length of a run of heads in tossing a fair coin?

10. The *variance* of a random variable X, denoted by σ^2, is defined as the mean value of the square of the deviations of the variable from its mean value, i.e.,

$$\sigma^2 = E([X - E(X)]^2).$$

(The nonnegative square root of the variance is called the *standard deviation* of X.) It may be shown[25] that

$$\sigma^2 = E(X^2) - [E(X)]^2$$

where, for a discrete random variable assuming integral values only, $E(X^2)$ is given by

$$E(X^2) = \sum_{k=0}^{\infty} k^2 P(X = k).$$

If Y is the generating function of the sequence $\{P(X = k)\}$, show that $E(X^2) = Y''(1) + Y'(1)$ (the primes indicate differentiations) and thus obtain the formula

$$\sigma^2 = Y''(1) + Y'(1) - [Y'(1)]^2.$$

11. The formula obtained in Problem 10 allows the calculation of the variance of a random variable X directly from the generating function of the probability distribution of X. Using the generating functions in (4.61) and in Problem 8,

---

[24] Generating functions are used to develop the theory of runs in Chap. 13 of Feller, cited in footnote 19. See also pp. 100–104 of Bush and Mosteller, cited in footnote 23. For a brief indication of the use of the theory of runs to test certain statistical hypotheses (e.g., testing for lack of randomness in sequences of observations) see P. G. Hoel, *Introduction to Mathematical Statistics*, 2nd ed., John Wiley, New York, 1954, pp. 293–299.

[25] See p. 213 of Feller, cited in footnote 19.

calculate the variance of X if (a) X is binomially distributed with parameters n and p, and (b) if X has the geometric distribution defined in Problem 8.

12. Reconsider the binomial probability model of the text (pp. 196–200) with the following modification: the probability of success is no longer constant from trial to trial but depends on the trial number. Let p_k denote the probability of success on trial number k. ($q_k = 1 - p_k$ is then the probability of failure on trial number k.) Suppose n independent trials are performed and let X and $y_{k,n}$ be defined as in the text. Show that the difference equation (4.56) and subsidiary conditions (4.57) and (4.58) become

$$y_{k+1,n+1} = q_{n+1}y_{k+1,n} + p_{n+1}y_{k,n} \qquad k = 0, 1, 2, \cdots, n; \ n = 0, 1, 2, \cdots$$

$$y_{0,n+1} = q_{n+1}y_{0,n}$$

$$y_{0,0} = 1 \qquad y_{k,0} = 0 \qquad \text{if } k > 0.$$

With the generating function Y_n defined as in (4.59), show that (4.61) now becomes

$$Y_n(s) = (q_1 + p_1 s)(q_2 + p_2 s) \cdots (q_n + p_n s) \qquad n = 1, 2, 3, \cdots$$

with $Y_0(s) = 1$ as before. Finally, use this generating function to show that

$$E(X) = p_1 + p_2 + \cdots + p_n.$$

(Note that all of these results reduce to the corresponding formulas in the binomial probability model of the text in the special case when $p_1 = p_2 = \cdots = p_n = p$.)

13. Let z be a function of two variables defined at all points (x, y) with non-negative integral coordinates. The value of z at (j, k) is denoted by $z_{j,k}$. In moving to the neighboring point $(j + 1, k + 1)$, we define a number of new functions: the *first partial difference*[26] *of z with respect to x*, denoted by $\underset{1}{\Delta} z$, whose value at (j, k) is given by

$$\underset{1}{\Delta} z_{j,k} = z_{j+1,k} - z_{j,k},$$

and the *first partial difference of z with respect to y*, denoted by $\underset{2}{\Delta} z$, whose value at (j, k) is given by

$$\underset{2}{\Delta} z_{j,k} = z_{j,k+1} - z_{j,k}.$$

Four *second partial differences* of z may be defined:

$$\underset{1}{\Delta}(\underset{1}{\Delta} z), \qquad \underset{2}{\Delta}(\underset{2}{\Delta} z), \qquad \underset{2}{\Delta}(\underset{1}{\Delta} z), \qquad \underset{1}{\Delta}(\underset{2}{\Delta} z).$$

---

[26] Partial difference operators and partial difference equations are discussed in Chaps. X and XII of Jordan, cited in footnote 17, and in L. M. Milne-Thomson, *The Calculus of Finite Differences*, Macmillan, London, 1933, pp. 423–429, 475–476.

Calculate the values of these four second partial differences at (j, k) and in particular note that

$$\underset{2}{\Delta}(\underset{1}{\Delta}z) = \underset{1}{\Delta}(\underset{2}{\Delta}z),$$

i.e., the order of differencing is immaterial.

14. Show that the partial difference equation (4.56) can be written as

$$\underset{2}{\Delta}y_{k+1,n} + p\underset{1}{\Delta}y_{k,n} = 0.$$

15. Show that (4.65) can be written as

$$\underset{2}{\Delta}p_{k+1,n} + \underset{1}{\Delta}(\tau_k p_{k,n}) = 0.$$

16. Verify the equivalence stated in the text between (4.73) and (4.74).

17. Show directly from (4.71) that the generating function $P_k(s)$ has a factor s^k and thus deduce that

$$p_{k,n} = 0 \qquad \text{if } n < k.$$

(This merely expresses the fact that a word can be recalled only once per trial so the number of recalls is at most equal to the number of trials.)

18. In the Miller-McGill model of verbal learning suppose we assume

$$\tau_k = 1 - (1 - a)^{k+1} \qquad k = 0, 1, 2, \cdots.$$

Show that

$$p_{0,n} = (1 - a)^n \qquad p_{1,n} = (1 - a)^{n-1}[1 - (1 - a)^n].$$

[Cf. formula (14) of the Miller-McGill paper where the general solution $p_{k,n}$ is found.]

19. Let E_n denote the mean (or expected) number of times a word is recalled in the first n trials. Show from the definition (4.63) that

$$E_n = \sum_{k=0}^{n} k p_{k,n}.$$

If ρ_{n+1} denotes the theoretically expected recall score (i.e., mean *proportion* of words recalled) on trial number $n + 1$, deduce that

$$\rho_{n+1} = E_{n+1} - E_n \qquad n = 0, 1, 2, \cdots.$$

By using these relations and the difference equation (4.65), show that

$$\rho_{n+1} = \sum_{k=0}^{n} \tau_k p_{k,n}.$$

(Cf. Miller and McGill, pp. 371–372.)

4.5 MATRIX METHODS

We first present a brief outline of matrix algebra. Only those definitions and results (many without proofs) are given that are needed for the matrix analysis of such systems of difference equations as we shall study later. Additional material may be found in the many textbooks on matrix theory.[27]

A *matrix* is a rectangular array of numbers indicated as follows:

$$(4.79) \qquad A = \begin{pmatrix} a_{11} & a_{12} & \cdots & a_{1n} \\ a_{21} & a_{22} & \cdots & a_{2n} \\ \cdot & \cdot & & \cdot \\ \cdot & \cdot & & \cdot \\ \cdot & \cdot & & \cdot \\ a_{m1} & a_{m2} & \cdots & a_{mn} \end{pmatrix}.$$

A is an $m \times n$ (to be read "m by n") matrix, i.e., it has m rows and n columns, and thus is made up of mn numbers. Each of these numbers is called an *element* of the matrix. The two subscripts on each element specify the row and the column (always in that order) in which this element is to be found. For example, a_{12} is the number appearing in the first row and second column; the matrix element in the ith row and jth column is a_{ij}. For an $m \times n$ matrix, the subscripts i and j range from 1 to m and from 1 to n respectively, and the matrix A is often written in the more space-conserving form

$$A = (a_{ij}) \qquad i = 1, 2, \cdots, m; \; j = 1, 2, \cdots, n.$$

The following are examples of matrices:

$$(4.80) \qquad A = \begin{pmatrix} 1 & 3 \\ -2 & 0 \end{pmatrix} \qquad B = \begin{pmatrix} 5 \\ -1 \end{pmatrix}$$

$$C = (3 \quad -6 \quad \tfrac{1}{2} \quad -1) \qquad D = \begin{pmatrix} 1 & 0 & 0 \\ 0 & -5 & 0 \\ 0 & 0 & 3 \end{pmatrix}.$$

A is a *square matrix* with 2 rows and 2 columns. (A square matrix is of *order* n if it has n rows and n columns.) B is a matrix with 2 rows but just 1 column. It is an example of a *column vector*. (Any $m \times 1$ matrix is said to be a column vector with m components.) C is a matrix with 1 row and 4 columns. It is a *row vector* with 4 components. (A $1 \times n$ matrix is called a row vector with n components.) Finally, D is a 3×3 *diagonal matrix*. [A square matrix is diagonal if all the elements off the main (upper left to lower right) diagonal are 0.]

[27] Of particular interest to students of the social sciences are the presentations in J. G. Kemeny, J. L. Snell, and G. L. Thompson, *Introduction to Finite Mathematics*, Prentice Hall, New York, 1957, especially Chaps. 5 and 7; L. L. Thurstone, *Multiple-Factor Analysis*, Univ. of Chicago Press, Chicago, 1947, pp. 1–50; R. G. D. Allen, *Mathematical Economics*, Macmillan, London, 1956, Chaps. 12–14.

Although a matrix is not a number, it is nevertheless possible to construct an algebra of matrices just as one constructs an algebra of numbers. To do this we must specify rules for combining matrices in various ways, i.e., by addition, multiplication, etc. We shall define and illustrate the following concepts of matrix algebra: (1) equality of two matrices, (2) addition of matrices, (3) multiplication of a matrix by a number, (4) multiplication of a matrix by a matrix, and (5) inverse (or reciprocal) of a matrix.

Two $m \times n$ matrices are said to be equal if each element of the first is equal to the corresponding element of the second. Letting

(4.81) $A = (a_{ij})$, $B = (b_{ij})$, $i = 1, 2, \cdots, m; \ j = 1, 2, \cdots, n$,

this means $A = B$ if and only if

(4.82) $a_{ij} = b_{ij}$ $i = 1, 2, \cdots, m; j = 1, 2, \cdots, n$.

It is important to note that the single matrix equation $A = B$ is equivalent to the mn equations specified in (4.82). For example, numbers x, y, z, and w which satisfy the matrix equation

$$\begin{pmatrix} x + y & y + z \\ z + w & w + x \end{pmatrix} = \begin{pmatrix} 1 & 0 \\ 0 & 1 \end{pmatrix}$$

must simultaneously satisfy all four algebraic equations

$$x + y = 1, \qquad y + z = 0, \qquad z + w = 0, \qquad w + x = 1.$$

Addition of matrices, as equality, is defined only for matrices of the same order, i.e., those having the same number of rows and columns. *The sum of two $m \times n$ matrices is defined as the $m \times n$ matrix obtained by adding corresponding elements.* Thus, with A and B as in (4.81), if

$$C = A + B = (c_{ij}) \qquad i = 1, 2, \cdots, m; \ j = 1, 2, \cdots, n,$$

then by definition,

$$c_{ij} = a_{ij} + b_{ij} \qquad i = 1, 2, \cdots, m; \ j = 1, 2, \cdots, n.$$

For example,

$$\begin{pmatrix} 4 & -1 \\ 2 & \frac{1}{2} \end{pmatrix} + \begin{pmatrix} -1 & 1 \\ 3 & \frac{1}{2} \end{pmatrix} = \begin{pmatrix} 3 & 0 \\ 5 & 1 \end{pmatrix}$$

and the sum of two column vectors is again a column vector:

$$\begin{pmatrix} x \\ y \end{pmatrix} + \begin{pmatrix} 1 \\ -1 \end{pmatrix} = \begin{pmatrix} x + 1 \\ y - 1 \end{pmatrix}.$$

If A is any matrix and c any number, the matrix obtained by multiplying every element of the matrix A by c is called the (scalar) multiple of A by c and is denoted by cA or Ac. For example, with the matrices A and B in (4.80), we have

$$2A = \begin{pmatrix} 2 & 6 \\ -4 & 0 \end{pmatrix} \qquad (-1)B = -B = \begin{pmatrix} -5 \\ 1 \end{pmatrix}.$$

If A and B are any two $m \times n$ matrices, then we define

$$A - B = A + (-1)B,$$

so that *the difference $A - B$ is obtained by subtracting from each element of A the corresponding element of B.*

With these definitions of scalar multiplication and addition of matrices, it is possible to prove the following important theorems of matrix algebra. These are obvious counterparts of (and as may be expected, their proofs depend upon) the analogous results in the algebra of numbers.

Commutative Law of Addition. If A and B are any two $m \times n$ matrices, then

$$A + B = B + A.$$

Associative Law of Addition. If A, B, and C are any three $m \times n$ matrices, then

$$A + (B + C) = (A + B) + C.$$

Distributive Law. If c_1 and c_2 are any numbers and A and B any $m \times n$ matrices, then

$$(c_1 + c_2)A = c_1A + c_2A,$$
$$c_1(A + B) = c_1A + c_1B.$$

Existence of a Zero. If A is any $m \times n$ matrix and if O is the $m \times n$ matrix whose elements are all zero, then

$$A + O = O + A = A.$$

(Note that although there is only one zero in the algebra of numbers, in matrix algebra there are infinitely many zero matrices, one for each pair of integers m and n.)

As a preliminary to defining the product of two matrices, consider first the special case in which R is a row vector and C a column vector, each with n components:

$$R = (x_1 \quad x_2 \quad \cdots \quad x_n), \qquad C = \begin{pmatrix} y_1 \\ y_2 \\ \cdot \\ \cdot \\ \cdot \\ y_n \end{pmatrix}.$$

The product of the row vector R and the column vector C, denoted by RC, is defined to be the 1×1 *matrix (which is identified with the corresponding number)*

$$(4.83) \qquad RC = \sum_{k=1}^{n} x_k y_k = x_1 y_1 + x_2 y_2 + \cdots + x_n y_n.$$

For example,

$$(5 \quad -1 \quad 2)\begin{pmatrix} 6 \\ 10 \\ -5 \end{pmatrix} = (5)(6) + (-1)(10) + (2)(-5) = 10.$$

Now if A is any $m \times n$ matrix, we may schematically write

$$(4.84) \quad A = (a_{ij}) = \begin{pmatrix} R_1 \\ R_2 \\ \cdot \\ \cdot \\ \cdot \\ R_m \end{pmatrix} \qquad i = 1, 2, \cdots, m; \; j = 1, 2, \cdots n,$$

where R_i is the $1 \times n$ row vector forming the ith row of the matrix A,

$$(4.85) \qquad R_i = (a_{i1} \quad a_{i2} \quad \cdots \quad a_{in}) \qquad i = 1, 2, \cdots, m.$$

Similarly, if B is any $n \times p$ matrix, we may write

$$(4.86) \qquad B = (b_{ij}) = (C_1 \quad C_2 \quad \cdots \quad C_p)$$
$$i = 1, 2, \cdots, n; \; j = 1, 2, \cdots, p,$$

where the jth column of the matrix B is the column vector

$$(4.87) \qquad C_j = \begin{pmatrix} b_{1j} \\ b_{2j} \\ \cdot \\ \cdot \\ \cdot \\ b_{nj} \end{pmatrix} = 1, 2, \cdots, p.$$

The product $R_i C_j$ has been defined in (4.83) and is thus given by

$$(4.88) \qquad R_i C_j = \sum_{k=1}^{n} a_{ik} b_{kj} = a_{i1} b_{1j} + a_{i2} b_{2j} + \cdots + a_{in} b_{nj},$$

where, as indicated in (4.85) and (4.87), the subscript i varies from 1 to m and the subscript j from 1 to p.

The product of the two matrices A and B in (4.84) *and* (4.86) *is defined as the matrix*

(4.89)
$$AB = \begin{pmatrix} R_1C_1 & R_1C_2 & \cdots & R_1C_p \\ R_2C_1 & R_2C_2 & \cdots & R_2C_p \\ \cdot & \cdot & & \cdot \\ \cdot & \cdot & & \cdot \\ \cdot & \cdot & & \cdot \\ R_mC_1 & R_mC_2 & \cdots & R_mC_p \end{pmatrix}.$$

Thus, *the element in the* ith *row and* jth *column of AB is the product of the* ith *row of A and the* jth *column of B, this product being* R_iC_j *as given in* (4.88).

The following matrix multiplications can be performed using this definition:

$$\begin{pmatrix} 1 & 4 \\ -1 & 5 \end{pmatrix} \begin{pmatrix} 3 \\ 1 \end{pmatrix} = \begin{pmatrix} 1 \cdot 3 + 4 \cdot 1 \\ -1 \cdot 3 + 5 \cdot 1 \end{pmatrix} = \begin{pmatrix} 7 \\ 2 \end{pmatrix}$$

(4.90)

$$\begin{pmatrix} 1 & -1 \\ 1 & 0 \end{pmatrix} \begin{pmatrix} 1 & 2 \\ 3 & 4 \end{pmatrix} = \begin{pmatrix} 1 \cdot 1 + -1 \cdot 3 & 1 \cdot 2 + -1 \cdot 4 \\ 1 \cdot 1 + 0 \cdot 3 & 1 \cdot 2 + 0 \cdot 4 \end{pmatrix} = \begin{pmatrix} -2 & -2 \\ 1 & 2 \end{pmatrix}$$

$$\begin{pmatrix} 1 & 2 \\ 3 & 4 \end{pmatrix} \begin{pmatrix} 1 & -1 \\ 1 & 0 \end{pmatrix} = \begin{pmatrix} 1 \cdot 1 + 2 \cdot 1 & 1 \cdot -1 + 2 \cdot 0 \\ 3 \cdot 1 + 4 \cdot 1 & 3 \cdot -1 + 4 \cdot 0 \end{pmatrix} = \begin{pmatrix} 3 & -1 \\ 7 & -3 \end{pmatrix}.$$

According to this definition of matrix multiplication, to form the product AB it is necessary that the number of columns in A be equal to the number of rows in B. If A is an $m \times n$ matrix and B an $n \times p$ matrix, then AB is an $m \times p$ matrix.

It is important to observe that matrix multiplication, unlike ordinary algebraic multiplication of numbers, is not commutative. *The product AB is in general not equal to BA.* In fact, recalling the remark made in the preceding paragraph, it is entirely possible for AB to be defined without BA being defined. This happens, for example, if A is a 2 × 4 and B a 4 × 3 matrix. Then AB is a 2 × 3 matrix but BA is undefined since one cannot multiply a matrix having 3 columns by one having 2 rows. But even when both AB and BA are defined, it may still happen that $AB \neq BA$. The last two matrix multiplications in (4.90) illustrate this noncommutativity.

Matrix multiplication does share the following properties of number multiplication:

Associative Law of Multiplication. If A, B, and C are any three $m \times n$, $n \times p$, and $p \times q$ matrices respectively, then

$$A(BC) = (AB)C.$$

Distributive Law. If A, B, and C are any three $m \times n$, $n \times p$, and $n \times p$ matrices respectively, then

$$A(B + C) = AB + AC.$$

Existence of a Unit Matrix. Let the $n \times n$ unit matrix I_n be defined for each positive integer n as the $n \times n$ diagonal matrix whose diagonal elements are all equal to 1, i.e.,

$$I_n = \begin{pmatrix} 1 & 0 & \cdots & 0 \\ 0 & 1 & \cdots & 0 \\ \cdot & \cdot & & \cdot \\ \cdot & \cdot & & \cdot \\ \cdot & \cdot & & \cdot \\ 0 & 0 & \cdots & 1 \end{pmatrix}.$$

Then if A is any $n \times n$ matrix,

$$AI_n = I_n A = A.$$

(The $n \times n$ unit matrix I_n is ordinarily written as I, the particular value of the subscript n being inferred from the context.)

The associative law enables us to define positive integral powers of matrices. The square matrix A being given, we write A^2 for the product AA. (Note that this product is defined if and only if A is square.) Since by the associative law

$$A^2 A = (AA)A = A(AA) = A(A^2),$$

i.e., A^2 multiplied by A or A multiplied by A^2 gives the same result, we denote this product by A^3. Continuing in this same manner, we observe that the product

$$AA \cdots A \qquad (n \text{ times})$$

is independent of the manner in which the matrices may be bracketed for multiplication and this product can therefore legitimately be denoted by A^n. With this definition, the laws of exponents in ordinary algebra carry over to matrix algebra:

$$A^m A^n = A^{m+n}, \qquad (A^m)^n = A^{mn},$$

where m and n are any positive integers.

In the algebra of real numbers, the reciprocal (or inverse) of a number a is defined as a number b, if it exists, such that $ab = ba = 1$. All numbers have reciprocals with the single exception of the number 0 and the reciprocal of a if $a \neq 0$ is that (unique) number denoted by a^{-1} or $1/a$.

A similar procedure is followed in matrix algebra. *If A is any matrix, then a matrix B, if it exists, such that*

$$AB = BA = I,$$

where I is a unit matrix, is called an inverse of A.

For both the products AB and BA to be defined and equal, it is necessary that both A and B be square matrices of the same order. Hence nonsquare matrices cannot possess an inverse.

But not even all square matrices have inverses. With each square matrix A of order n,

$$A = \begin{pmatrix} a_{11} & a_{12} & \cdots & a_{1n} \\ a_{21} & a_{22} & \cdots & a_{2n} \\ \cdot & \cdot & & \cdot \\ \cdot & \cdot & & \cdot \\ \cdot & \cdot & & \cdot \\ a_{n1} & a_{n2} & \cdots & a_{nn} \end{pmatrix},$$

there is associated a *number* called the *determinant* of A and denoted by $|A|$,

$$|A| = \begin{vmatrix} a_{11} & a_{12} & \cdots & a_{1n} \\ a_{21} & a_{22} & \cdots & a_{2n} \\ \cdot & \cdot & & \cdot \\ \cdot & \cdot & & \cdot \\ \cdot & \cdot & & \cdot \\ a_{n1} & a_{n2} & \cdots & a_{nn} \end{vmatrix}.$$

This is the determinant which the reader has met before in algebra and in this book (p. 130). If A is a 2×2 matrix,

$$A = \begin{pmatrix} a_{11} & a_{12} \\ a_{21} & a_{22} \end{pmatrix},$$

the determinant of the matrix A is the number

$$|A| = a_{11}a_{22} - a_{12}a_{21}.$$

(Formulas for the determinants of higher order matrices can also be written, as in footnote 12, p. 163, but we shall not need them.) Thus, if

(4.91) $$A = \begin{pmatrix} 5 & 1 \\ 2 & 4 \end{pmatrix}, \qquad B = \begin{pmatrix} 1 & 2 \\ 2 & 4 \end{pmatrix},$$

then

$$|A| = (5)(4) - (2)(1) = 18, \qquad |B| = (1)(4) - (2)(2) = 0.$$

A square matrix A is said to be singular if $|A| = 0$, nonsingular if $|A| \neq 0$. It is possible to prove that the necessary and sufficient condition for a square matrix A to possess an inverse is that $|A| \neq 0$. Hence, only nonsingular matrices possess inverses and all singular matrices in matrix algebra share the fate of 0 in ordinary algebra in not having reciprocals or inverses. It can further be proved that a nonsingular matrix A has a *unique* inverse. This inverse is denoted by A^{-1} so that

$$AA^{-1} = A^{-1}A = I.$$

The reader can verify that the nonsingular matrix A in (4.91) has an inverse given by

$$A^{-1} = \begin{pmatrix} \frac{2}{9} & -\frac{1}{18} \\ -\frac{1}{9} & \frac{5}{18} \end{pmatrix}.$$

and as a further exercise can show that the matrix I is its own inverse (just as 1 is its own reciprocal in number algebra). Since B in (4.91) is a singular matrix, it has no inverse.

The three examples that follow, although an interruption of the main mathematical development of this section, serve two purposes. Leading as they do to systems of simultaneous difference equations, they enable us to formulate such systems using matrix algebra, and thus illustrate the relevance of the algebra of matrices to the subject matter of this book. And perhaps more important, our interest in solving such systems of difference equations will lead us to the consideration of some important, more advanced aspects of matrix theory.

Example 1

Neural networks have been the subject of much recent research,[28] both by psychologists interested in studying the animal nervous system and by mathematicians and engineers exploring the potentialities of large-scale electronic computing machines. Consider the network from a to b (Figure 4.2) which may be considered a rudimentary memory unit. With time divided into intervals of equal length (approximately the duration of synaptic delay), we assume neuron b fires only in the interval following that in which neuron a transmits an impulse to b. If a is once stimulated, say at $t = 0$, then one impulse reaches b and fires it at $t = 1$, while another returns to a at $t = 1$, thus restimulating a to send another impulse causing b to fire at $t = 2$. Simultaneously, the cycle is completed by restimulating neuron a at $t = 2$, which fires b at $t = 3$, etc. Thus the original stimulus to a is remembered, as it were, and b is fired in each

---

[28] See, e.g., A. S. Householder, "Neural Structure in Perception and Response," *Psychological Review*, *54* (1947), 169–176, and the articles in C. E. Shannon and J. McCarthy (eds.), *Automata Studies*, Annals of Mathematics Studies, No. 34, Princeton Univ. Press, Princeton, 1956.

later time interval. According to Householder (p. 174), "physiological learning might consist in the activation of such cycles · · ·."

But now suppose[29] that this memory unit has a certain fallibility which allows the possibility of neuron a not restimulating itself when it is once fired or of firing spontaneously in the absence of any stimulus. Suppose we denote by ε the probability of making either type of error. It is

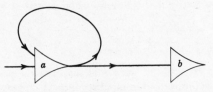

Figure 4.2

convenient to think of this unit as being in one of two states at any time t (with $t = 0, 1, 2, \cdot\cdot\cdot$): state 1 when neuron a is stimulated and thus fires b (at time $t + 1$) and state 2 when a is inactive. In each time interval (i.e., in going from any time t to time $t + 1$) neuron a may continue in its existing state or, because of the allowed possibility of error, may alter its state. The following transitions may take place with the indicated probabilities:

From State	to State	with Probability
1	→	1	$1 - \varepsilon$
1	→	2	ε
2	→	1	ε
2	→	2	$1 - \varepsilon$

If we assume that initially (at $t = 0$) the unit is in state 1 (i.e., neuron a transmits an impulse to b), then we are interested in the state of the unit at subsequent times. If $\varepsilon = 0$, the memory unit works without error and we have already seen that it remains in state 1 for all time. But with $\varepsilon > 0$, we are unable to say with certainty whether neuron a will or will not fire. For example, if $\varepsilon = 0.01$, then in the long run neuron a will transmit another impulse to b in 99 and be inactive in 1 out of every 100 time intervals following one in which it is in state 1.

Let p_t = probability that unit is in state 1 at time t, and q_t = probability that unit is in state 2 at time t, where $t = 0, 1, 2, \cdot\cdot\cdot$. (Of course, $q_t = 1 - p_t$ so that q_t is easily computed as soon as one finds p_t. But

[29] J. von Neumann, "Probabilistic Logics and the Synthesis of Reliable Organisms from Unreliable Components," p. 62 of Shannon and McCarthy, cited in footnote 28.

it turns out to be more convenient for our matrix analysis to write one difference equation for *each* of the two unknown functions p and q.)

The unit can be in state 1 at time $t + 1$ only by being in state 1 at time t. and then undergoing the transition $1 \rightarrow 1$, or by being in state 2 at time t and moving from state 2 to state 1 in the time interval from t to $t + 1$. Hence

$$(4.92) \qquad p_{t+1} = p_t(1 - \varepsilon) + q_t\varepsilon \qquad t = 0, 1, 2, \cdots.$$

A similar analysis of the possible ways of being in state 2 at time $t + 1$ leads to the corresponding equation

$$(4.93) \qquad q_{t+1} = p_t\varepsilon + q_t(1 - \varepsilon) \qquad t = 0, 1, 2, \cdots.$$

Equations (4.92) and (4.93) together constitute a pair of difference equations and we must find functions p and q which satisfy both. With the initial conditions

$$(4.94) \qquad\qquad p_0 = 1 \qquad q_0 = 0$$

expressing the fact that neuron a is stimulated at time $t = 0$, we shall apply matrix methods to reformulate and then solve this system of difference equations.

Example 2

In *panel surveys* a sample of individuals is selected from the population under study and the members of this panel are interviewed on repeated occasions. The basic data obtained are the successive responses of the same individuals to a series of questions. Such surveys are therefore especially well suited for gathering information not only on percentages of responses in different categories at different times but also on percentages of individuals who change in successive interviews from one response to another. It is, for example, a matter of some interest to know not only that a panel is evenly split on a given question on two successive interviews but also whether the second 50–50 division is due to all respondents maintaining their previous answers or to the fact that an equal number of individuals reversed themselves.

Consider[30] an individual on such a panel confronted with a question allowing only one of the two answers "yes" or "no." With respect to

---

[30] T. W. Anderson, "Probability Models for Analyzing Time Changes in Attitudes," Chap. 1 in P. F. Lazarsfeld (ed.), *Mathematical Thinking in the Social Sciences*, Free Press, Glencoe, 1954. This paper discusses much more general mathematical models than those presented here and also applies the analysis to data actually obtained in the panel survey of voters reported in P. F. Lazarsfeld, B. Berelson, and H. Gaudet, *The People's Choice*, Columbia Univ. Press, New York, 1948.

this particular question, the respondent may be classified into one of two states: state 1 if his answer is "yes" and state 2 if his answer is "no." The same question being asked of this panel member in a succeeding interview, he may again be placed in either state 1 or state 2. An individual may maintain or change his answer from one interview to the next. Hence the following transitions may take place with the indicated probabilities:

From State	to State	with Probability
1	$\rightarrow$ 1	$1 - \alpha$
1	$\rightarrow$ 2	α
2	$\rightarrow$ 1	β
2	$\rightarrow$ 2	$1 - \beta$

For the sake of generality, we have introduced the symbols α and β to denote the probabilities of an individual changing his answer from "yes" to "no" and from "no" to "yes" respectively. Of course, $0 \leq \alpha \leq 1, 0 \leq \beta \leq 1$, and the probabilities with which the panel member maintains his "yes" and "no" response in two successive interviews are $1 - \alpha$ and $1 - \beta$ respectively.

Now if we know the initial opinion (with respect to the question under discussion) of an individual in the first panel interview (denoted by $t = 0$) and if we assume his subsequent answers governed by the aforementioned transition probabilities, then we should be able to calculate the probability of any sequence of opinions held in later panel interviews (denoted by $t = 1, 2, 3, \cdots$). A sample computation follows in the special case where the panel is evenly divided at $t = 0$ with $\alpha = 0.2$ and $\beta = 0.3$. Then in the next interview, i.e., at $t = 1$, of the 50% who said "yes," $1 - \alpha = 0.8$ or 80% again say "yes" whereas $\alpha = 0.2$ or 20% change their minds and say "no." Hence $(80)(50) = 40\%$ of the total panel are characterized by the sequence of answers "yes" at $t = 0$, "yes" at $t = 1$, whereas $(20)(50) = 10\%$ follow the sequence "yes" at $t = 0$, "no" at $t = 1$. Similarly, we find $(30)(50) = 15\%$ saying "no" at $t = 0$, "yes" at $t = 1$, and $(70)(50) = 35\%$ saying "no" at $t = 0$, "no" at $t = 1$. Thus in the second interview, the original 50–50 split has changed to a yes-no percentage of 55–45. Proceeding in the same way, this calculation may be continued for interviews number 2, 3, 4, $\cdots$ to find the yes-no percentages 57.5–42.5, 58.75–41.25, 59.375–40.625, $\cdots$. These appear to be approaching a 60–40 split and those limiting proportions will actually be shown to be correct when the analysis in the general case is performed.

For this purpose we let p_t denote the probability that an individual is in state 1 (i.e., answers "yes") in panel interview number t. Similarly,

$q_t = 1 - p_t$ denotes the probability of being in state 2 (i.e., answering "no") in this same interview.

An individual answering "yes" in interview number $t + 1$ either answered "yes" in interview number t and then maintained his opinion in the next interview or answered "no" in interview number t and then reversed himself in the following interview. Hence

(4.95) $$p_{t+1} = p_t(1 - \alpha) + q_t\beta \qquad t = 0, 1, 2, \cdots.$$

A similar analysis of an individual answering "no" in interview number $t + 1$ leads to the corresponding equation

(4.96) $$q_{t+1} = p_t\alpha + q_t(1 - \beta) \qquad t = 0, 1, 2, \cdots.$$

Here, as in (4.92) and (4.93), we have a system of two simultaneous difference equations for the two unknown functions p and q. We shall see that these equations, together with the prescribed values p_0 and q_0 (which specify the yes-no split in the initial panel interview), will determine p_t and q_t and therefore the yes-no split in interview number t for all subsequent t-values.

Example 3

In certain approaches to statistical theories of learning,[31] with any given organism there is associated a set S^* whose elements are all the stimulus events (visual, auditory, etc.) that may occur in the organism in any situation whatever. On any experimental trial, some of these elements of S^* may occur and others may fail to occur. Behavior on a given trial is assumed to be a function of the stimulus elements which are sampled on that trial.

On this basis, Estes[32] has constructed a theory to account for changes in response tendencies (e.g., spontaneous recovery, forgetting) during time intervals when the organism and the stimulus situation are well separated. One aspect of this work is a theory of stimulus fluctuation which we consider here. During any trial run in a given experiment, not all of the stimulus elements in S^* are present. In any time period the complete set S^* is, in fact, subdivided into two sets: the subset S of stimulus elements which are available during that time period and the complementary set S' of elements not available. In this theory, the probability of a response is equal to the proportion of elements in the available set S

---

[31] See the discussion and references in Chap. 2 of Bush and Mosteller, cited in footnote 23.

[32] W. K. Estes, "Statistical Theory of Spontaneous Recovery and Regression," *Psychological Review*, 62 (1955), 145–154.

that are conditioned to that response and it is therefore of some interest to study the fluctuations in membership of the two sets S and S'.

Time is divided into periods of equal length Δt, with an initial experimental period $t = 0$ followed by t-values 1, 2, 3, $\cdots$ denoting later times Δt, $2\Delta t$, $3\Delta t$, $\cdots$. Concentrating on a specified element in the total set S^*, let us say that this element is in state 1 at any time t if it is available (i.e., in S) and in state 2 if it is not available (i.e., in S'). Because of environmental fluctuations, in any time interval Δt a stimulus element may either maintain or alter its state, i.e., "there is some probability j that an element in the available set S will become unavailable, i.e., go into S', during any given interval Δt, and a probability j' that an element in S' will enter S." (Estes, p. 147.)

Hence the following transitions may take place (in any time interval of length Δt) with the indicated probabilities:

From State	to State	with Probability
1	$\rightarrow$	1	$1 - j$
1	$\rightarrow$	2	j
2	$\rightarrow$	1	j'
2	$\rightarrow$	2	$1 - j'$

We want to study the functions p and q defined as follows: $p_t = $ probability that a given element of S^* is in state 1 (i.e., available) at time t, and $q_t = $ probability that this element is in state 2 (i.e., unavailable) at time t. Although it is clear that $q_t = 1 - p_t$, we again prefer to prepare the way for the matrix analysis to follow by treating p and q as equally important and writing a difference equation for each.

The given element is in state 1 at time $t + 1$ if and only if one of the following mutually exclusive events occurs: (1) it was in state 1 at time t and stayed there in the time interval from t to $t + 1$, or (2) it was in state 2 at time t and moved into state 1 in this time interval. Hence

$$(4.97) \qquad p_{t+1} = p_t(1 - j) + q_t j' \qquad t = 0, 1, 2, \cdots.$$

A similar analysis of the possibilities for being in state 2 at $t + 1$ leads to the equation

$$(4.98) \qquad q_{t+1} = p_t j + q_t(1 - j') \qquad t = 0, 1, 2, \cdots.$$

We shall show, using matrix methods, that with the initial probabilities p_0 and q_0 prescribed, the system of two difference equations (4.97) and (4.98) uniquely determines the probabilities p_t and q_t for all t-values. With these probabilities known, it is possible (see Estes, p. 151) to derive expressions for the proportions of conditioned elements, and therefore the response probabilities, in S and S' respectively, at time t following an experimental period.   And these expressions yield theoretical curves of spontaneous recovery and regression capable of empirical testing.

It should be clear that these three examples have identical mathematical structures and differ only in the interpretations of the symbols used.   All fall within the following general framework: a sequence of trials, each with only two possible outcomes, takes place at times $t = 0, 1, 2, \cdots$. It is customary to introduce a new terminology and say that the system is in state 1 or state 2 at any time t according as the trial taking place at that time results in one or the other of the two possible outcomes.   The initial probabilities p_0 and $q_0 (= 1 - p_0)$ for the system to be in state 1 and state 2, respectively, at $t = 0$ are prescribed, as are the constant transition probabilities for the system to pass from any state to any state from one trial to the next, i.e., in the time interval between consecutive t-values:

$$P(1 \to 1) = 1 - \alpha \qquad P(1 \to 2) = \alpha$$
$$P(2 \to 1) = \beta \qquad P(2 \to 2) = 1 - \beta.$$

Here α and β are any numbers between 0 and 1 inclusive, and $P(j \to k)$ denotes the probability of the transition from state j to state k.

With p_t and q_t denoting the probabilities of finding the system in state 1 and state 2, respectively, at time t, we have seen in each of the examples that we are led to the system of difference equations

$$(4.99) \qquad \begin{aligned} p_{t+1} &= (1 - \alpha)p_t + \beta q_t \\ q_{t+1} &= \alpha p_t + (1 - \beta)q_t \end{aligned} \qquad t = 0, 1, 2, \cdots.$$

We want to find expressions for p_t and q_t in terms of the given initial probabilities p_0 and q_0, the probabilities α and β, and the time t.   That is, we want to find the probability distribution of the state of the system at any time t from the prescribed data: the probability distribution of the initial state of the system at $t = 0$. the transition probabilities governing the random fluctuations in the state of the system from trial to trial, and the elapsed time from $t = 0$ (or the number of trials which have taken place).

A sequence of trials of the type just described is called a *Markov chain*. For the sake of simplicity, we have considered Markov chains with only two states, but the general theory is concerned with trials having any number of possible outcomes.   For example, in the panel survey model of Example 2 it is actually more realistic to allow respondents to answer "don't know" in addition to "yes" and "no," and this leads to a Markov chain with three states.

The study[33] of finite Markov chains is an important part of modern probability theory and has many applications. But we shall not dwell on the probability aspects of this subject since our main interest is in the systems of difference equations which arise in such chains.

The system of two simultaneous difference equations (4.99) can be written as a single matrix difference equation. We let V_t denote the column vector with two components p_t and q_t, i.e.,

$$(4.100) \qquad V_t = \begin{pmatrix} p_t \\ q_t \end{pmatrix} \qquad t = 0, 1, 2, \cdots.$$

If we denote by A the matrix of transition probabilities

$$(4.101) \qquad A = \begin{pmatrix} 1 - \alpha & \beta \\ \alpha & 1 - \beta \end{pmatrix},$$

then the system (4.99) may be written

$$(4.102) \qquad V_{t+1} = A V_t \qquad t = 0, 1, 2, \cdots.$$

That this matrix equation is actually equivalent to the two equations in (4.99) follows by use of the rules for equality and multiplication of matrices. For (4.102) is shorthand for

$$\begin{pmatrix} p_{t+1} \\ q_{t+1} \end{pmatrix} = \begin{pmatrix} 1 - \alpha & \beta \\ \alpha & 1 - \beta \end{pmatrix} \begin{pmatrix} p_t \\ q_t \end{pmatrix},$$

which upon performing the matrix multiplication becomes

$$\begin{pmatrix} p_{t+1} \\ q_{t+1} \end{pmatrix} = \begin{pmatrix} (1 - \alpha)p_t + \beta q_t \\ \alpha p_t + (1 - \beta)q_t \end{pmatrix}.$$

But this is equivalent to (4.99) since two matrices are equal if and only if their corresponding elements are equal.

The matrix

$$(4.103) \qquad V_0 = \begin{pmatrix} p_0 \\ q_0 \end{pmatrix}$$

being prescribed (since p_0 and q_0 are given), the matrix difference equation (4.102) enables us to calculate V_t for any value of t. For, putting $t = 0$ in (4.102) we obtain

$$(4.104) \qquad V_1 = A V_0.$$

[33] See Chaps. 15 and 16 in W. Feller, *An Introduction to Probability Theory and Its Applications*, Vol. 1, 2nd ed., John Wiley, New York, 1957. Feller uses the method of generating functions in his algebraic treatment of probabilities in finite Markov chains. Comparable results using matrix theory may be found in M. Frechet, *Recherches théoriques modernes sur le calcul des probabilités*, Vol. 2 (Théorie des événements en chaine dans le cas d'un nombre finit d'états possibles), Gauthier-Villars, Paris, 1938.

Now putting $t = 1$ in (4.102),

$V_2 = AV_1 = A(AV_0)$, by (4.104),

$\quad = (AA)V_0$, by the associative law for matrix multiplication,

$\quad = A^2 V_0$.

Similarly, we find

$$V_3 = AV_2 = A(A^2 V_0) = (AA^2)V_0 = A^3 V_0,$$

and it is not difficult to show (by mathematical induction) that

(4.105) $\qquad V_t = A^t V_0 \qquad t = 0, 1, 2, \cdots.$

[For (4.105) to be correct, i.e., reduce to V_0, when $t = 0$ requires the agreement that $A^0 = I$, the identity or unit matrix.]

The problem of solving the system of difference equations (4.99) has now been reduced to the problem of finding the tth power of matrix A. For with A^t known, we can perform the matrix multiplication $A^t V_0$ and thus, by (4.105), find V_t. And the two components of the column vector V_t are just the desired probabilities p_t and q_t.

Unfortunately, the elements of the tth power of an arbitrary matrix, and in particular matrix A, are not easy to find directly. But the powers of a *diagonal* matrix are very easy to obtain. For if

(4.106) $\qquad D = \begin{pmatrix} \lambda_1 & 0 \\ 0 & \lambda_2 \end{pmatrix},$

then, by the definition of matrix multiplication, we calculate

$$D^2 = \begin{pmatrix} \lambda_1^2 & 0 \\ 0 & \lambda_2^2 \end{pmatrix}, \qquad D^3 = \begin{pmatrix} \lambda_1^3 & 0 \\ 0 & \lambda_2^3 \end{pmatrix},$$

and, in general, D^t is the diagonal matrix obtained by raising each diagonal element of D to the tth power:

(4.107) $\qquad D^t = \begin{pmatrix} \lambda_1^t & 0 \\ 0 & \lambda_2^t \end{pmatrix} \qquad t = 1, 2, 3, \cdots.$

In view of the ease with which D^t may be found, it is of some practical importance to investigate the possibility of reducing the problem of calculating A^t to that of calculating D^t for some appropriate choice of the diagonal matrix D. For this purpose we introduce the notion of similar matrices.

A square matrix A is said to be similar to another matrix B if there exists a nonsingular matrix M such that

(4.108) $\qquad B = M^{-1}AM.$

One reason for the importance of similar matrices (and this accounts for our interest) is the fact that if (4.108) holds, then

$$(4.109) \qquad B^t = M^{-1}A^tM \qquad t = 1, 2, 3, \cdots;$$

i.e., the very same matrix M guaranteed by the hypothesis that A is similar to B also makes A^t similar to B^t for all positive integral powers t. To prove that (4.109) follows from (4.108) we note that

$$\begin{aligned} B^2 &= (M^{-1}AM)(M^{-1}AM), \\ &= M^{-1}A(MM^{-1})AM, \text{ by associativity} \\ &= M^{-1}AIAM, \text{ since } MM^{-1} = I, \\ &= M^{-1}A^2M, \text{ since } AIA = A^2. \end{aligned}$$

Similarly,

$$\begin{aligned} B^3 = B^2B &= (M^{-1}A^2M)(M^{-1}AM) \\ &= M^{-1}A^2(MM^{-1})AM \\ &= M^{-1}A^2IAM \\ &= M^{-1}A^3M, \end{aligned}$$

and a formal proof that (4.109) is true for all indicated t-values may be completed using mathematical induction.

We are interested in finding A^t for matrix A in (4.101). Suppose A is similar to some matrix B. Then (4.108) and hence (4.109) hold. From (4.109) we obtain, by premultiplication by M,

$$MB^t = M(M^{-1}A^tM) = (MM^{-1})A^tM = IA^tM = A^tM,$$

and then by postmultiplication by M^{-1},

$$MB^tM^{-1} = (A^tM)M^{-1} = A^t(MM^{-1}) = A^tI = A^t,$$

or, in summary,

$$(4.110) \qquad A^t = MB^tM^{-1} \qquad t = 0, 1, 2, \cdots.$$

[Since matrix multiplication is not commutative, it is crucial to distinguish between premultiplication and postmultiplication. In the argument leading to (4.110) we perform first one, then the other, in order to twice obtain the simplifying combination $MM^{-1} = I$.]

Thus, if A is similar to B, (4.110) shows that the calculation of A^t may be replaced by the calculation of B^t followed by two matrix multiplications (premultiplication by M and postmultiplication by M^{-1}) to obtain MB^tM^{-1}.

Since powers of diagonal matrices are trivial to compute, it is natural to inquire whether A is similar to some diagonal matrix D.   For then,

(4.111) $$D = M^{-1}AM$$

and, by (4.110),

(4.112) $$A^t = MD^tM^{-1} \qquad t = 0, 1, 2, \cdots,$$

so that A^t may be computed fairly easily.

Suppose, for the moment, that an arbitrary 2×2 matrix A is similar to some diagonal matrix D with diagonal elements λ_1 and λ_2.   From (4.111) there follows by premultiplication by M,

$$AM = MD.$$

If we denote by C_1 and C_2 the two column vectors of the 2×2 matrix M, then

(4.113) $$M = (C_1 \quad C_2)$$

and

(4.114) $$A(C_1 \quad C_2) = (C_1 \quad C_2) \begin{pmatrix} \lambda_1 & 0 \\ 0 & \lambda_2 \end{pmatrix}.$$

But by the definition of matrix multiplication,

$$A(C_1 \quad C_2) = (AC_1 \quad AC_2),$$

and

$$(C_1 \quad C_2) \begin{pmatrix} \lambda_1 & 0 \\ 0 & \lambda_2 \end{pmatrix} = (\lambda_1 C_1 \quad \lambda_2 C_2).$$

Hence (4.114) becomes

$$(AC_1 \quad AC_2) = (\lambda_1 C_1 \quad \lambda_2 C_2)$$

which, by the definition of equality of matrices, leads to the two equations

(4.115) $$AC_1 = \lambda_1 C_1 \qquad AC_2 = \lambda_2 C_2.$$

These equations enable us to introduce two important concepts in matrix theory.   *Any nonzero column vector X is said to be a characteristic vector of the matrix A if there exists a number λ such that*

(4.116) $$AX = \lambda X.$$

The number λ is called a characteristic root of A corresponding to the characteristic vector X and vice versa.   (Characteristic vectors are also called eigenvectors and characteristic roots are often called latent roots or eigenvalues.)

From (4.115) it follows that if (4.111) holds, i.e., *if A is similar to a diagonal matrix D, then the columns of matrix M are characteristic vectors of A and the diagonal elements of D are the corresponding characteristic roots of A.*

Let us apply these results to matrix A in (4.101). To find the characteristic roots and vectors of A we must solve the matrix equation $AX = \lambda X$ or, letting x_1 and x_2 denote the components of the column vector X,

$$\begin{pmatrix} 1 - \alpha & \beta \\ \alpha & 1 - \beta \end{pmatrix} \begin{pmatrix} x_1 \\ x_2 \end{pmatrix} = \lambda \begin{pmatrix} x_1 \\ x_2 \end{pmatrix}.$$

Performing the indicated matrix and scalar multiplication, we have

$$\begin{pmatrix} (1 - \alpha)x_1 + \beta x_2 \\ \alpha x_1 + (1 - \beta)x_2 \end{pmatrix} = \begin{pmatrix} \lambda x_1 \\ \lambda x_2 \end{pmatrix}.$$

By the definition of equality of matrices,

$$(1 - \alpha)x_1 + \beta x_2 = \lambda x_1$$
$$\alpha x_1 + (1 - \beta)x_2 = \lambda x_2,$$

and transposing terms, we obtain

(4.117)
$$(1 - \alpha - \lambda)x_1 + \beta x_2 = 0$$
$$\alpha x_1 + (1 - \beta - \lambda)x_2 = 0.$$

The equations in (4.117) form a set of two simultaneous linear equations in the two unknowns x_1 and x_2. We require a nontrivial solution (i.e., not both $x_1 = 0$ and $x_2 = 0$) since, by definition, a characteristic vector must be a nonzero vector. In order to have a nontrivial solution of the system (4.117) it is necessary and sufficient for the determinant formed by the coefficients to be zero[34]:

(4.118)
$$\begin{vmatrix} 1 - \alpha - \lambda & \beta \\ \alpha & 1 - \beta - \lambda \end{vmatrix} = 0.$$

Equation (4.118) is known as the *characteristic equation* of matrix A. Its roots are the characteristic roots of A. (In general, for any square matrix A, if the determinant of the matrix $A - \lambda I$ is denoted by $|A - \lambda I|$, then the equation

(4.119)
$$|A - \lambda I| = 0$$

is called the characteristic equation of matrix A. If A is $n \times n$, then this equation is a polynomial of the nth degree in the unknown λ.)

[34] For a proof see L. E. Dickson, *New First Course in the Theory of Equations*, John Wiley, New York, 1939, p. 131. This result is actually equivalent to Theorem 3.4.

Expanding (4.118) we find

$$(1 - \alpha - \lambda)(1 - \beta - \lambda) - \alpha\beta = 0$$

or

$$\lambda^2 - (2 - \alpha - \beta)\lambda + (1 - \alpha - \beta) = 0.$$

This quadratic equation for λ factors into

$$(\lambda - 1)[\lambda - (1 - \alpha - \beta)] = 0$$

so we have the two characteristic roots

(4.120) $$\lambda_1 = 1 \qquad \lambda_2 = 1 - \alpha - \beta.$$

When $\lambda = \lambda_1 = 1$, both equations in (4.117) become

(4.121) $$-\alpha x_1 + \beta x_2 = 0$$

so that x_1 and x_2 are not uniquely determined (there being one equation for two unknowns), but are determined up to a multiplicative constant. [This is a consequence of the fact that if (4.116) holds, then also $A(kX) = \lambda(kX)$ for arbitrary constant k. Thus, if X is a characteristic vector of A corresponding to the characteristic root λ, then so is kX.]

We choose as a solution to (4.121)

$$x_1 = \beta \qquad x_2 = \alpha,$$

and letting C_1 denote the characteristic vector corresponding to $\lambda = \lambda_1 = 1$,

(4.122) $$C_1 = \begin{pmatrix} \beta \\ \alpha \end{pmatrix}.$$

Since we require nonzero characteristic vectors, we insist that α and β are not both zero. [The limiting case $\alpha = \beta = 0$ is of no interest since the Markov chain then allows no transitions from one state to the other, the difference equations (4.99) become trivial, and there is no need for any matrix analysis.]

When $\lambda = \lambda_2 = 1 - \alpha - \beta$, the equations (4.117) reduce to

$$\beta x_1 + \beta x_2 = 0 \qquad \alpha x_1 + \alpha x_2 = 0,$$

which are satisfied if we choose

$$x_1 = 1 \qquad x_2 = -1.$$

Letting C_2 denote the characteristic vector corresponding to $\lambda = \lambda_2 = 1 - \alpha - \beta$,

(4.123) $$C_2 = \begin{pmatrix} 1 \\ -1 \end{pmatrix}.$$

If $\alpha + \beta \neq 0$, the two characteristic roots λ_1 and λ_2 of matrix A are unequal. In this case, it may be shown[35] that A is indeed similar to a diagonal matrix, i.e., (4.111) and hence (4.112) hold. From our previous discussion it follows that the elements of this diagonal matrix D are the characteristic roots $\lambda_1 = 1$ and $\lambda_2 = 1 - \alpha - \beta$ of A and the columns of matrix M are the corresponding vectors C_1 and C_2 in (4.122) and (4.123). Hence (4.112) becomes

$$(4.124) \qquad A^t = M \begin{pmatrix} 1 & 0 \\ 0 & (1 - \alpha - \beta)^t \end{pmatrix} M^{-1} \qquad t = 0, 1, 2, \cdots,$$

where

$$(4.125) \qquad M = (C_1 \quad C_2) = \begin{pmatrix} \beta & 1 \\ \alpha & -1 \end{pmatrix}.$$

To complete the calculation of A^t, the inverse M^{-1} of matrix M must be found and then the matrix multiplications indicated in (4.124) can be performed. We leave the details of these computations for the problems, merely state the results here, and then consider a much simpler method for computing A^t.

One finds the inverse of M is

$$(4.126) \qquad\qquad M^{-1} = \frac{1}{\alpha + \beta} \begin{pmatrix} 1 & 1 \\ \alpha & -\beta \end{pmatrix},$$

so that, performing the multiplications required in (4.124),

$$(4.127) \qquad A^t = \frac{1}{\alpha + \beta} \begin{pmatrix} \beta & \beta \\ \alpha & \alpha \end{pmatrix} + \frac{(1 - \alpha - \beta)^t}{\alpha + \beta} \begin{pmatrix} \alpha & -\beta \\ -\alpha & \beta \end{pmatrix}.$$

Hence, recalling (4.105) (and the fact that $p_0 + q_0 = 1$),

$$(4.128) \qquad V_t = A^t V_0 = \frac{1}{\alpha + \beta} \begin{pmatrix} \beta \\ \alpha \end{pmatrix} + (1 - \alpha - \beta)^t \begin{pmatrix} p_0 - \dfrac{\beta}{\alpha + \beta} \\ q_0 - \dfrac{\alpha}{\alpha + \beta} \end{pmatrix}.$$

Since the components of V_t are p_t and q_t, we find the following solution of the system of difference equations (4.99):

$$(4.129) \qquad \begin{aligned} p_t &= \frac{\beta}{\alpha + \beta} + (1 - \alpha - \beta)^t \left(p_0 - \frac{\beta}{\alpha + \beta} \right) \\ q_t &= \frac{\alpha}{\alpha + \beta} + (1 - \alpha - \beta)^t \left(q_0 - \frac{\alpha}{\alpha + \beta} \right) \end{aligned} \qquad t = 0, 1, 2, \cdots.$$

[35] For a proof that every square matrix A with distinct characteristic roots is similar to a diagonal matrix see G. Birkhoff and S. MacLane, *A Survey of Modern Algebra*, rev. ed., Macmillan, New York, 1953, p. 327.

Before interpreting this solution for each of the three examples with which this section began, we pause to present a much simpler method of finding A^t. This method depends on the fundamental *Cayley-Hamilton theorem*:[36] *Every square matrix satisfies its characteristic equation.* This means that if each power λ^i in the characteristic equation of the $n \times n$ matrix A,

$$(4.130) \qquad f(\lambda) = |A - \lambda I| = a_0 + a_1\lambda + a_2\lambda^2 + \cdots + a_n\lambda^n = 0,$$

is replaced by the same power A^i of the matrix (and if $a_0 = a_0\lambda^0$ is replaced by $a_0 A^0 = a_0 I$), the result is the zero matrix, i.e.,

$$(4.131) \qquad f(A) = a_0 I + a_1 A + a_2 A^2 + \cdots + a_n A^n = O.$$

To illustrate this result, consider the matrix

$$A = \begin{pmatrix} 3 & 2 \\ 4 & 1 \end{pmatrix}$$

whose characteristic equation is

$$|A - \lambda I| = \begin{vmatrix} 3 - \lambda & 2 \\ 4 & 1 - \lambda \end{vmatrix} = (3 - \lambda)(1 - \lambda) - 8 = \lambda^2 - 4\lambda - 5 = 0.$$

The Cayley-Hamilton theorem asserts that

$$A^2 - 4A - 5I = O,$$

a fact which may be verified by first computing

$$A^2 = AA = \begin{pmatrix} 17 & 8 \\ 16 & 9 \end{pmatrix},$$

and then noting that

$$A^2 - 4A - 5I = \begin{pmatrix} 17 & 8 \\ 16 & 9 \end{pmatrix} - \begin{pmatrix} 12 & 8 \\ 16 & 4 \end{pmatrix} - \begin{pmatrix} 5 & 0 \\ 0 & 5 \end{pmatrix} = \begin{pmatrix} 0 & 0 \\ 0 & 0 \end{pmatrix} = O.$$

For the argument to follow, we shall assume A is a 2×2 matrix. (The general case of an $n \times n$ matrix is treated in precisely the same way; see 10 of Problems 4.5.) As we have seen, with A a 2×2 matrix, its characteristic equation is a quadratic equation, i.e., $n = 2$ in (4.130) and

$$(4.132) \qquad f(\lambda) = a_0 + a_1\lambda + a_2\lambda^2.$$

Now suppose λ^t is divided by $f(\lambda)$. We obtain a quotient polynomial $q(\lambda)$ and a remainder polynomial $r(\lambda)$, i.e.,

$$(4.133) \qquad \lambda^t = f(\lambda)q(\lambda) + r(\lambda),$$

[36] See p. 320 of Birkhoff and MacLane, cited in footnote 35, for a proof.

and since we are dividing by a quadratic polynomial, the remainder $r(\lambda)$ is at most of first degree. Thus constants a and b exist such that

$$(4.134) \qquad\qquad r(\lambda) = a + b\lambda.$$

(The situation here is entirely analogous to what happens when dividing two numbers. For example, when 13 is divided by 5, we obtain a quotient 2 and remainder 3, i.e., $13 = 5 \cdot 2 + 3$. Of course, when dividing by 5, we are certain that the remainder is at most 4.)

Since (4.133) is an identity in λ, we may[37] replace λ by matrix A and obtain

$$A^t = f(A)q(A) + r(A).$$

But, by the Cayley-Hamilton theorem $f(A) = O$. Hence

$$(4.135) \qquad\qquad A^t = r(A) = aI + bA.$$

The problem of finding the power A^t is thus reduced to identifying the coefficients a and b in (4.135). To do this, we use the fact that the roots of the characteristic equation $f(\lambda) = 0$ are the characteristic roots of matrix A. If these roots are denoted by λ_1 and λ_2, then

$$(4.136) \qquad\qquad f(\lambda_1) = 0 \qquad f(\lambda_2) = 0,$$

and from (4.133) and (4.134),

$$(4.137) \qquad \lambda_1{}^t = r(\lambda_1) = a + b\lambda_1 \qquad \lambda_2{}^t = r(\lambda_2) = a + b\lambda_2.$$

The determinant of this system of two simultaneous linear equations for the unknowns a and b,

$$\begin{vmatrix} 1 & \lambda_1 \\ 1 & \lambda_2 \end{vmatrix} = \lambda_2 - \lambda_1,$$

is different from zero if the characteristic roots are distinct. This is precisely the condition (see Theorem 3.4) which allows a and b to be uniquely determined from the two equations in (4.137). We summarize in the following rule:

RULE FOR FINDING A^t: *Let A be a 2×2 matrix with distinct[38] characteristic roots λ_1 and λ_2. These roots are found by solving the quadratic equation*

$$(4.138) \qquad\qquad |A - \lambda I| = 0.$$

---

[37] This requires proof. See the very brief mention in Birkhoff and MacLane, p. 316 cited in footnote 35, and in R. R. Stoll, *Linear Algebra and Matrix Theory*, McGraw-Hill, New York, 1952, pp. 163–164.

[38] The adjustments required in the case of $n \times n$ matrices and equal characteristic roots are considered in 10 and 11 of Problems 4.5.

Determine the constants a and b by solving the simultaneous system of linear equations (4.137). *Then*

$$(4.139) \qquad\qquad A^t = aI + bA.$$

Let us use this rule to calculate A^t for the matrix (4.101) and thus to verify the expression in (4.127). The characteristic equation of A, given by (4.118), has the two characteristic roots $\lambda_1 = 1$ and $\lambda_2 = 1 - \alpha - \beta$. Since we have assumed $\alpha + \beta \neq 0$, these roots are unequal and the rule applies. The system of linear equations (4.137) becomes

$$1 = a + b \qquad (1 - \alpha - \beta)^t = a + b(1 - \alpha - \beta),$$

which is found to have the solution

$$a = \frac{(1 - \alpha - \beta)^t - (1 - \alpha - \beta)}{\alpha + \beta} \qquad b = \frac{1 - (1 - \alpha - \beta)^t}{\alpha + \beta}$$

Hence, by substitution in (4.139),

$$
\begin{aligned}
A^t &= a \begin{pmatrix} 1 & 0 \\ 0 & 1 \end{pmatrix} + b \begin{pmatrix} 1 - \alpha & \beta \\ \alpha & 1 - \beta \end{pmatrix} \\[2mm]
&= \begin{pmatrix} a + b - b\alpha & b\beta \\ b\alpha & a + b - b\beta \end{pmatrix} \\[2mm]
&= \frac{1}{\alpha + \beta} \begin{pmatrix} \beta + \alpha(1 - \alpha - \beta)^t & \beta - \beta(1 - \alpha - \beta)^t \\ \alpha - \alpha(1 - \alpha - \beta)^t & \alpha + \beta(1 - \alpha - \beta)^t \end{pmatrix} \\[2mm]
&= \frac{1}{\alpha + \beta} \begin{pmatrix} \beta & \beta \\ \alpha & \alpha \end{pmatrix} + \frac{(1 - \alpha - \beta)^t}{\alpha + \beta} \begin{pmatrix} \alpha & -\beta \\ -\alpha & \beta \end{pmatrix},
\end{aligned}
$$

as in (4.127).

This example illustrates the advantage of using the rule rather than (4.112) to calculate A^t. Once having calculated the characteristic roots, the use of the rule eliminates the calculation of matrix M and its inverse as well as the matrix multiplications involved in (4.112).

Now we return to the system of difference equations (4.99) which we rewrite here for easy reference:

$$(4.140) \qquad \begin{aligned} p_{t+1} &= (1 - \alpha)p_t + \beta q_t \\ q_{t+1} &= \alpha p_t + (1 - \beta)q_t \end{aligned} \qquad t = 0, 1, 2, \cdots.$$

Recall that α and β are transition probabilities so that $0 \leq \alpha \leq 1$ and $0 \leq \beta \leq 1$. We first consider the two limiting cases (1) $\alpha = 0$ and $\beta = 0$, and (2) $\alpha = 1$ and $\beta = 1$.

If $\alpha = 0$ and $\beta = 0$, then (4.140) reduces to

$$p_{t+1} = p_t \qquad q_{t+1} = q_t \qquad t = 0, 1, 2, \cdots,$$

from which it follows that

$$p_t = p_0 \qquad q_t = q_0 \qquad t = 0, 1, 2, \cdots.$$

This means that no transitions from any state to any other state are possible in this Markov chain. If the system is initially in state 1, for example (i.e., if $p_0 = 1$, $q_0 = 0$), then $p_t = 1$ and the system remains in state 1 for all time.

The second case, $\alpha = 1$ and $\beta = 1$, is of only slightly greater interest. Now (4.140) becomes

$$p_{t+1} = q_t \qquad q_{t+1} = p_t \qquad t = 0, 1, 2, \cdots,$$

from which we find

$$p_{t+2} = q_{t+1} = p_t \qquad q_{t+2} = p_{t+1} = q_t \qquad t = 0, 1, 2, \cdots.$$

It follows (by putting $t = 0, 1, 2, \cdots$ in turn) that

$$p_2 = p_4 = p_6 = \cdots = p_0 \qquad p_1 = p_3 = p_5 = \cdots = q_0 = 1 - p_0.$$

Thus, if the system is initially in state 1 (i.e., $p_0 = 1$, $q_0 = 0$), then the system moves to state 2 at $t = 1$ (since $p_1 = 0$), returns to state 1 at $t = 2$ (since $p_2 = 1$), and continues to alternate between its two states. We have a Markov chain in which transitions from one state to the other are certain in each time interval.

The case where $0 < \alpha + \beta < 2$ remains to be considered. For this purpose return to the solutions (4.129) of the system of difference equations (4.140). We repeat p_t here:

$$(4.141) \quad p_t = \frac{\beta}{\alpha + \beta} + (1 - \alpha - \beta)^t \left(p_0 - \frac{\beta}{\alpha + \beta} \right) \qquad t = 0, 1, 2, \cdots.$$

Two questions are of special interest: (1) Is there a limiting value of the probability p_t as time increases, i.e., does $\lim_{t \to \infty} p_t$ exist? (2) Assuming this limiting value does exist, what is its dependence on the prescribed initial values p_0 and q_0?

In the case under consideration, the quantity $1 - \alpha - \beta$ is a number between -1 and $+1$ and $(1 - \alpha - \beta)^t$ therefore approaches the limiting value 0 as $t \to \infty$. Hence we find

$$(4.142) \qquad\qquad p_t \to \frac{\beta}{\alpha + \beta} \qquad \text{as } t \to \infty.$$

As t increases without bound, i.e., as more and more trials are performed, the probability p_t of finding the system in state 1 approaches the value $\beta/(\alpha + \beta)$. *This limiting value is clearly independent of p_0 and q_0 and therefore is not influenced by the initial state of the system at $t = 0$.* We interpret this result for each of the three examples which began this section.

Example 1 (continued from p. 215). Here $\alpha = \beta = \varepsilon$ so that $p_t \to \frac{1}{2}$ as $t \to \infty$. Therefore, "after a long time, the memory content of the machine disappears, since it tends to equal likelihood of being right or wrong, i.e., to irrelevancy."[39]

Example 2 (continued from p. 217). Here p_t, the probability of a respondent answering "yes" in poll number t, approaches a value independent of p_0, his initial probability of answering "yes." "Thus, knowledge of the present opinion of the person is of less and less use in prediction as the time span increases. This confirms our intuition that our ability to predict a person's opinion decreases as the time interval increases."[40]

Example 3 (continued from p. 219). Following Estes, the probabilities α and β are denoted by j and j' respectively. Here p_t, the probability that a stimulus element is in the available set S at time t, "will settle down to a constant value J $[= j'/(j + j')]$ after a sufficiently long interval of time, and the total numbers of elements in S and S' will stabilize at mean values N and N', respectively, which satisfy the relation $N = J(N + N')$."[41] This last equation follows from the fact that, in the limit, a proportion J of the total number of elements $(N + N')$ will on the average belong to the available set S.

We conclude this section with some further examples of the use of matrix methods to solve systems of difference equations.

Example 4.

Suppose a function x, defined for t-values $0, 1, 2, \cdots$, satisfies the second-order difference equation

$$(4.143) \qquad x_{t+1} - 3x_t + 2x_{t-1} = 0 \qquad t = 1, 2, 3, \cdots,$$

with the prescribed initial values

$$x_0 = 0 \qquad x_1 = 1.$$

This is a homogeneous difference equation with constant coefficients and can be solved by the method of Chapter 3. Since the auxiliary

---

[39] See p. 62 of von Neumann, cited in footnote 29.

[40] See p. 30 of Anderson, cited in footnote 30.

[41] See p. 151 of Estes, cited in footnote 32.

equation $m^2 - 3m + 2 = 0$ has roots 1 and 2, the general solution is given by $x_t = A + B2^t$. Taking into consideration the initial values, we find $A = -1$, $B = 1$, so that

$$x_t = 2^t - 1 \qquad t = 0, 1, 2, \cdots.$$

We want to show that this second-order equation is equivalent to a system of two first-order difference equations and may therefore be solved using matrix methods. This is only a special instance of an important result pertaining to the transformation of a single high-order equation into a number of first-order equations.[42]

We introduce a new function y (defined only for t-values $1, 2, 3, \cdots$) in order to write the given difference equation as a system of two first-order equations:

$$(4.144) \qquad x_{t+1} = 3x_t - 2y_t \qquad y_{t+1} = x_t \qquad t = 1, 2, 3, \cdots.$$

If we define the column vectors

$$V_t = \begin{pmatrix} x_t \\ y_t \end{pmatrix} \qquad t = 1, 2, 3, \cdots,$$

the system (4.144) may be written in matrix form as

$$(4.145) \qquad V_{t+1} = AV_t \qquad t = 1, 2, 3, \cdots,$$

where A is the matrix formed by the coefficients, i.e.,

$$A = \begin{pmatrix} 3 & -2 \\ 1 & 0 \end{pmatrix}.$$

From (4.145) we obtain

$$V_{t+1} = AV_t = A^2 V_{t-1} = \cdots = A^t V_1 \qquad t = 0, 1, 2, \cdots,$$

which, as indicated, is correct for $t = 0$ as well as $t = 1, 2, 3, \cdots$. The vector V_1 is known since

$$V_1 = \begin{pmatrix} x_1 \\ y_1 \end{pmatrix} = \begin{pmatrix} x_1 \\ x_0 \end{pmatrix} = \begin{pmatrix} 1 \\ 0 \end{pmatrix}.$$

To find A^t we use the rule on p. 230. The characteristic equation is

$$\begin{vmatrix} 3 - \lambda & -2 \\ 1 & -\lambda \end{vmatrix} = \lambda^2 - 3\lambda + 2 = (\lambda - 1)(\lambda - 2) = 0,$$

---

[42] See the discussions in P. A. Samuelson, *Foundations of Economic Analysis*, Harvard Univ. Press, 1948, pp. 383–388, and B. Friedman, *Principles and Techniques of Applied Mathematics*, John Wiley, New York, 1956, pp. 122–123.

so the characteristic roots of A are $\lambda_1 = 1$ and $\lambda_2 = 2$. Hence

$$A^t = aI + bA$$

and the constant coefficients a and b are determined from the simultaneous equations

$$1 = a + b \qquad 2^t = a + 2b.$$

Solving this system we find $a = 2 - 2^t$ and $b = 2^t - 1$ so that

$$A^t = (2 - 2^t) \begin{pmatrix} 1 & 0 \\ 0 & 1 \end{pmatrix} + (2^t - 1) \begin{pmatrix} 3 & -2 \\ 1 & 0 \end{pmatrix}.$$

Finally, we calculate

$$V_{t+1} = A^t V_1 = \begin{pmatrix} 2^{t+1} - 1 \\ 2^t - 1 \end{pmatrix} \qquad t = 0, 1, 2, \cdots.$$

Since $x_t = y_{t+1}$ is the second component of this column vector, we find the solution of (4.143) given by

$$x_t = 2^t - 1 \qquad t = 0, 1, 2, \cdots.$$

Example 5.

Consider the following foreign trade model[43] where, for simplicity, we treat the special case of two countries.   We assume

(i) National income (Y) equals consumption outlays (C) plus net investment (i) plus exports (X) minus imports (M).

(ii) Outlays for domestic consumption (D) equal total consumption (C) minus imports (M).

(iii) Time is divided into periods of equal length, denoted by $t = 0, 1, 2, \cdots$, so, for example, $Y(t)$ denotes the national income in period t.   All quantities vary from period to period with the exception of net investment which is assumed to maintain a constant level.

We use subscripts 1 and 2 to distinguish the two countries.   Thus $Y_1(t)$ is the national income of country 1 in period t and $X_2(t)$ denotes the exports of country 2 in that same period.

---

[43] For the economics background and generalization to any finite number of countries, see L. A. Metzler, "Underemployment Equilibrium in International Trade," *Econometrica*, *10* (1942), 97–112, and "A Multiple-Region Theory of Income and Trade," *Econometrica*, *18* (1950), 329–354; R. M. Goodwin, "The Multiplier as Matrix," *Economic Journal*, *59* (1949), 537–555; J. S. Chipman, "Professor Goodwin's Matrix Multiplier," *Economic Journal*, *60* (1950), 753–763, and "The Multi-Sector Multiplier," *Econometrica*, *18* (1950), 355–374.   See also F. Machlup, *International Trade and the National Income Multiplier*, Blakiston Co., Philadelphia, 1943, where, in Appendix A, the technique of Chapter 3, rather than matrix methods, is used to develop a formula giving income increments at any time after an autonomous increase in export in a foreign trade model involving only two countries.

Assumptions (i) and (ii), when written for country 1 in period t, become

$$Y_1(t) = C_1(t) + i_1(t) + X_1(t) - M_1(t),$$
$$D_1(t) = C_1(t) - M_1(t).$$

With the aid of the second of these equations, the first can be written

(4.146) $$Y_1(t) = D_1(t) + X_1(t) + i_1,$$

where we have written i_1 for the constant net investment.

We obtain a corresponding equation when assumptions (i) and (ii) are applied to country 2:

(4.147) $$Y_2(t) = D_2(t) + X_2(t) + i_2.$$

Now domestic consumption as well as the level of imports are reasonably assumed to depend on the national income level. In fact, we suppose that this dependence is of the following simple type:

(iv) Domestic consumption (D) and imports (M) of each country in any period are constant multiples of the country's national income (Y) one time period earlier. In symbols,

$$D_1(t) = m_{11} Y_1(t-1) \qquad M_1(t) = m_{21} Y_1(t-1)$$
$$D_2(t) = m_{22} Y_2(t-1) \qquad M_2(t) = m_{12} Y_2(t-1),$$

where $m_{11}, m_{21}, m_{22}, m_{12}$ are constants (called marginal propensities).

Since we consider only two countries, the exports of one must be the imports of the other, i.e.,

$$M_1(t) = X_2(t) \qquad M_2(t) = X_1(t).$$

Inserting these equations and those resulting from assumption (iv), (4.146) and (4.147) become

(4.148)
$$Y_1(t) = m_{11} Y_1(t-1) + m_{12} Y_2(t-1) + i_1$$
$$Y_2(t) = m_{21} Y_1(t-1) + m_{22} Y_2(t-1) + i_2.$$

This system of difference equations may be written in matrix form as

$$\begin{pmatrix} Y_1(t) \\ Y_2(t) \end{pmatrix} = \begin{pmatrix} m_{11} & m_{12} \\ m_{21} & m_{22} \end{pmatrix} \begin{pmatrix} Y_1(t-1) \\ Y_2(t-1) \end{pmatrix} + \begin{pmatrix} i_1 \\ i_2 \end{pmatrix}$$

or, with the obvious definitions of matrices Y, M (the matrix multiplier), and i,

(4.149) $$Y(t) = M Y(t-1) + i \qquad t = 1, 2, 3, \cdots.$$

We assume that Y_1 and Y_2 are given for $t = 0$ so that $Y(0)$ is prescribed.

Putting $t = 1, 2, \cdots$ in turn in (4.149), we find

$$Y(1) = M Y(0) + i,$$
$$Y(2) = M Y(1) + i = M[M Y(0) + i] + i = M^2 Y(0) + (M + I)i,$$

and, in general, as can be proved by mathematical induction,

$$(4.150) \qquad Y(t) = M^t Y(0) + (M^{t-1} + M^{t-2} + \cdots + I)i.$$

But, by performing the required matrix multiplications, we can show

$$(I + M + M^2 + \cdots + M^{t-1})(I - M) = I - M^t.$$

Thus, *if the matrix* $(I - M)$ *is nonsingular* [so that its inverse $(I - M)^{-1}$ exists],

$$(I + M + M^2 + \cdots + M^{t-1}) = (I - M^t)(I - M)^{-1}.$$

In this case, the solution (4.150) can be written as

$$(4.151) \qquad Y(t) = M^t Y(0) + (I - M^t)(I - M)^{-1}i \qquad t = 0, 1, 2, \cdots.$$

It follows that a necessary and sufficient condition for the national income matrix Y to approach a matrix of constant (equilibrium) values independently of the initial matrix $Y(0)$ is that $M^t \to O$ (i.e., each element of M^t approach 0) as $t \to \infty$. Then

$$(4.152) \qquad Y(t) \to (I - M)^{-1}i.$$

To discover conditions under which we may be sure that $M^t \to O$, assume that the characteristic values, say λ_1 and λ_2, of matrix M are distinct. Then (see footnote 35) M is similar to a diagonal matrix D whose diagonal elements are λ_1 and λ_2, i.e., a nonsingular matrix S exists for which $D = S^{-1}MS$. It follows [with obvious notational changes from (4.112)] that $M^t = SD^tS^{-1}$. Since the nonzero elements of D^t are λ_1^{t} and λ_2^{t}, it further follows that if λ_1 and λ_2 are less than 1 in absolute value, each element of D^t will approach 0 as $t \to \infty$ and hence $M^t \to O$. For this reason, the following result is of some interest:

THEOREM 4.4. *If the elements of matrix M are positive and if the sum of the elements in each column is less than 1, the characteristic roots of M are less than 1 in absolute value.*

PROOF. If λ is a characteristic root of M, then, as in (4.115), there is a nonzero column vector C with components c_1 and c_2 for which $\lambda C = MC$. By performing the multiplications in $\lambda C = MC$, we have

$$\lambda c_1 = m_{11}c_1 + m_{12}c_2$$
$$\lambda c_2 = m_{21}c_1 + m_{22}c_2.$$

Taking absolute values, we find

$$|\lambda c_1| = |m_{11}c_1 + m_{12}c_2| \qquad |\lambda c_2| = |m_{21}c_1 + m_{22}c_2|.$$

Now, using the facts that the absolute value of a product is the product of the absolute values of the factors and the absolute value of a sum is at most the sum of the absolute values of the individual terms, we have

$$|\lambda||c_1| \le |m_{11}c_1| + |m_{12}c_2| = m_{11}|c_1| + m_{12}|c_2|,$$
$$|\lambda||c_2| \le |m_{21}c_1| + |m_{22}c_2| = m_{21}|c_1| + m_{22}|c_2|.$$

Adding these inequalities and using the hypothesis, we obtain

$$|\lambda|(|c_1| + |c_2|) \le (m_{11} + m_{21})|c_1| + (m_{12} + m_{22})|c_2| < |c_1| + |c_2|.$$

Hence $|\lambda| < 1$, which was to be proved.

Note that if the characteristic roots of M are less than 1 in absolute value, matrix $(I - M)$ is nonsingular, as required in the argument leading to (4.151) and (4.152). (To prove this, observe that the characteristic equation $|M - \lambda I| = 0$ is not satisfied by $\lambda = 1$ and therefore the determinant $|M - I|$ is different from zero.)

We summarize: If every marginal propensity m_{ij} $(i, j = 1, 2)$ is positive, and if each country's marginal propensity to spend (for consumption and imports) is less than 1, i.e.,

$$m_{11} + m_{21} < 1 \qquad m_{12} + m_{22} < 1,$$

then the national incomes of countries 1 and 2 approach equilibrium values given by the components of the limit column vector in (4.152).

Example 6.

The social structure of the Natchez Indians has been studied in great detail in the anthropological literature. This example describes a simple mathematical formulation of the rules of marriage and descent in this society.[44]

The Natchez consist of two classes, the Honored and the Stinkards. Honored individuals are further subdivided into three groups: the Suns, the Nobles, and the Honored.

---

[44] The mathematical model presented here is due to R. R. Bush, who has kindly permitted the use of his unpublished discussion. For the anthropological background and other references, see C. W. M. Hart, "A Reconsideration of the Natchez Social Structure," *American Anthropologist*, New Series, *45* (1943), 374–386. For a different mathematical treatment of marriage rules, see A. Weil, "Sur l'étude algébrique de certains types de lois de mariage," Chap. XIV in C. Levi-Strauss, *Les Structures Élémentaires de la Parente*, Presses Universitaires de France, Paris, 1949.

Both marriage and descent rules can conveniently be summarized in Table 4.3. The entry in each cell gives the class of a child of parents in the corresponding margins; × indicates that the corresponding marriage is not allowed.

TABLE 4.3

Father

		Sun	Noble	Honored	Stinkard
	Sun	×	×	×	Sun
	Noble	×	×	×	Noble
Mother	Honored	×	×	×	Honored
	Stinkard	Noble	Honored	Stinkard	Stinkard

We see, for example, that Suns, Nobles, and Honoreds must marry Stinkards whereas Stinkards may marry in any class. Also, children whose mother is a Sun are Suns but children whose father is a Sun become Nobles.

The question of interest here (and this is the subject of Hart's paper cited in footnote 44) is whether it is possible to have a stable distribution, i.e., a situation in which each of the four classes maintains constant membership, under these rules of marriage and descent. The development to follow, with additional hypotheses to be stated shortly, will supply a negative answer to this question.

The assumptions to be made are: (i) each class has an equal number of men and women in each generation, (ii) each individual marries once and only once, and (iii) each married couple has exactly one son and one daughter.

Let $x_i(n)$ denote the number of men in class i in generation n, where the classes are numbered 1, 2, 3, 4 representing Suns, Nobles, Honoreds, and Stinkards respectively. By assumption (i), $x_i(n)$ is also the number of women in class i in generation n.

Every Sun mother has precisely one son and mothers of other classes cannot produce Sun children. Hence

$$(4.153) \qquad x_1(n + 1) = x_1(n) \qquad n = 0, 1, 2, \cdots.$$

A Noble son is produced by every Sun father and every Noble mother and by no other parents. Hence

$$(4.154) \qquad x_2(n + 1) = x_1(n) + x_2(n) \qquad n = 0, 1, 2, \cdots.$$

An Honored son is produced by every Noble father and every Honored mother and by no other parents. Hence

(4.155) $x_3(n + 1) = x_2(n) + x_3(n)$ $n = 0, 1, 2, \cdots$.

To obtain a difference equation for the number of Stinkards in generation $(n + 1)$, we note that, by hypothesis, the total number of men (and women) is the same in each generation. Hence

$$x_1(n + 1) + x_2(n + 1) + x_3(n + 1) + x_4(n + 1)$$
$$= x_1(n) + x_2(n) + x_3(n) + x_4(n).$$

By using (4.153), (4.154), and (4.155), we can simplify and find

(4.156) $x_4(n + 1) = -x_1(n) - x_2(n) + x_4(n)$ $n = 0, 1, 2, \cdots$.

The system of equations (4.153) through (4.156) can be put into matrix form. We let

(4.157) $X(n) = \begin{pmatrix} x_1(n) \\ x_2(n) \\ x_3(n) \\ x_4(n) \end{pmatrix}$ $A = \begin{pmatrix} 1 & 0 & 0 & 0 \\ 1 & 1 & 0 & 0 \\ 0 & 1 & 1 & 0 \\ -1 & -1 & 0 & 1 \end{pmatrix}$

and then can verify by performing the matrix multiplication that

(4.158) $X(n + 1) = A X(n)$ $n = 0, 1, 2, \cdots$.

Hence, we have

(4.159) $X(n) = A^n X(0)$ $n = 0, 1, 2, \cdots$

and are once again faced with the problem of calculating the powers of a matrix.

We employ an interesting (and often useful) trick. We write

$$A = I + B,$$

where I is the 4×4 unit matrix and

$$B = \begin{pmatrix} 0 & 0 & 0 & 0 \\ 1 & 0 & 0 & 0 \\ 0 & 1 & 0 & 0 \\ -1 & -1 & 0 & 0 \end{pmatrix}.$$

Because I and B commute, we may use the binomial expansion to get A^n in terms of powers of I and B. (See 1 (f) of Problems 4.5.)

(4.160) $A^n = (I + B)^n = I^n + nI^{n-1}B + \dfrac{n(n - 1)}{2} I^{n-2}B^2 + \cdots + B^n.$

What makes this method successful is the fact (easily verified by doing the multiplications) that

$$B^2 = \begin{pmatrix} 0 & 0 & 0 & 0 \\ 0 & 0 & 0 & 0 \\ 1 & 0 & 0 & 0 \\ -1 & 0 & 0 & 0 \end{pmatrix},$$

but B^3 and hence all higher powers of B equal the zero matrix. Since all powers of I are equal to I, (4.160) becomes

$$A^n = I + nB + \frac{n(n-1)}{2} B^2.$$

Multiplying this by the initial vector $X(0)$, we get the column vector

$$(4.161) \quad X(n) = A^n X(0) = \begin{pmatrix} x_1(0) \\ x_2(0) + nx_1(0) \\ x_3(0) + nx_2(0) + \dfrac{n(n-1)}{2} x_1(0) \\ x_4(0) - nx_2(0) - \dfrac{n(n+1)}{2} x_1(0) \end{pmatrix},$$

which gives the number of men (or women) in each of the four classes in any generation n.

From this solution one sees that if and only if $x_1(0) = x_2(0) = 0$, i.e., if Suns and Nobles are initially absent from the population, will there be a limiting stable distribution. But then the vector $X(n) = X(0)$ for every n and there are never any fluctuations in numbers among the various classes.

But with Suns and Nobles present, there is no stable distribution under the marriage and descent laws assumed here. The number of Stinkards decreases with n until there are an insufficient number of Stinkard men to marry all the Sun, Noble, and Honored women as required by the marriage rules. Thus the social system as described cannot persist.

PROBLEMS 4.5

1. Let

$$A = \begin{pmatrix} 1 & 2 \\ 3 & 2 \end{pmatrix} \quad B = \begin{pmatrix} -1 & 1 \\ 1 & 0 \end{pmatrix} \quad C = \begin{pmatrix} 1 \\ 5 \end{pmatrix} \quad D = \begin{pmatrix} 2 & 0 \\ 0 & 3 \end{pmatrix}.$$

(a) Calculate $A + B$, $B + D$, and $A + D$. Is $A + C$ defined?

(b) Calculate AB and BA. Are they equal?

(c) Calculate DA and AD and compare the results with A. Generalize by proving that premultiplication (postmultiplication) of any square matrix A by any diagonal matrix D multiplies each row (column) of A by the corresponding diagonal element of D.

(d) Calculate AC. Is CA defined?

(e) Show that $A(B + D) = AB + AD$.

(f) Calculate A^2 and B^2. Show that

$$(A + B)^2 = (A + B)(A + B) = A^2 + AB + BA + B^2,$$

but the usual formula of ordinary algebra is incorrect, i.e.,

$$(A + B)^2 \neq A^2 + 2AB + B^2.$$

Explain. Show that this formula is valid if $AB = BA$.

(g) Show that A is nonsingular. Calculate the inverse of A as follows: Let $A^{-1} = \begin{pmatrix} x & y \\ z & w \end{pmatrix}$ and write the four equations expressed by the matrix identity $AA^{-1} = I$. Solve to find $x = -\frac{1}{2}, y = \frac{1}{2}, z = \frac{3}{4}, w = -\frac{1}{4}$.

(h) Use the method of part (g) to show that $B^{-1} = \begin{pmatrix} 0 & 1 \\ 1 & 1 \end{pmatrix}$.

(i) Show that the characteristic roots of A are -1 and 4 and find corresponding characteristic vectors.

(j) Show that A is similar to a diagonal matrix by finding M and D such that $D = M^{-1}AM$.

(k) Use the result of part (j) and (4.112) to show that

$$A^t = \frac{4^t}{5} \begin{pmatrix} 2 & 2 \\ 3 & 3 \end{pmatrix} + \frac{(-1)^t}{5} \begin{pmatrix} 3 & -2 \\ -3 & 2 \end{pmatrix}.$$

(l) Use the rule on p. 230 to calculate A^t.

(m) Show that the system of difference equations

$$x_{t+1} = x_t + 2y_t \qquad y_{t+1} = 3x_t + 2y_t \qquad t = 0, 1, 2, \cdots$$

can be written in matrix form as

$$X_{t+1} = AX_t \qquad t = 0, 1, 2, \cdots.$$

What are the components of the column vector X_t? Show that $X_t = A^t X_0$ and then use the result of part (k) to find (assuming $x_0 = 1$ and $y_0 = 0$)

$$x_t = \tfrac{1}{5}[2(4)^t + 3(-1)^t] \qquad y_t = \tfrac{3}{5}[(4)^t - (-1)^t] \qquad t = 0, 1, 2, \cdots.$$

(n) Show that the characteristic roots of B are $(-1 \pm \sqrt{5})/2$.

(o) What are the characteristic roots of matrix D?

2. Use the method of Problem 1(g) to calculate the inverse M^{-1} given in (4.126).

3. Perform the calculations (omitted in the text) leading to (4.128) and (4.129).

4. Place each of the following in the context of the two-state Markov chain model described on pp. 221 ff. and illustrated by Examples 1–3, pages 215–221. Identify the meanings of states 1 and 2, probabilities p_t and q_t, and transition probabilities α and β in each case. Specialize the system of difference equations (4.99) to each case and use (4.142) to arrive at the limiting value of the probability of finding the system in state 1.

(a) Consider an experiment[45] in which only two alternative responses (called "right" and "wrong") are possible. In repetitions of this experiment, a right response is followed by another right response 97% of the time; wrong follows wrong 73% of the time. Assuming a wrong response on trial 0, show that the probability of a subject choosing a right response on trial number t is given by

$$p_t = 0.9 \, [1 - (0.7)^t] \qquad t = 0, 1, 2, \cdots.$$

(b) Each letter in the Samoan language is either a vowel or a consonant. A consonant never follows a consonant but a vowel is followed by a consonant 49% of the time.[46]

(c) Suppose each day's weather is classified as "good" or "bad." Assume that the probability is 0.9 that the weather on an arbitrary day will be of the same kind as on the preceding day.

5. Solve the following second-order difference equations by replacing each by a system of two first-order equations and then following the procedure illustrated in Example 4 of the text.

(a) $x_{t+1} - 4x_t + 3x_{t-1} = 0 \qquad t = 1, 2, 3, \cdots$
   $x_0 = 0, x_1 = 1.$

(b) $x_{t+1} - x_t - 2x_{t-1} = 0 \qquad t = 1, 2, 3, \cdots$
   $x_0 = 1, x_1 = 0.$

(c) $x_{t+1} - x_{t-1} = 0 \qquad t = 1, 2, 3, \cdots$
   $x_0 = 1, x_1 = 3.$

6. Show that the system of three difference equations

$$x_{n+1} = 3x_n + 5y_n + 2z_n$$
$$y_{n+1} = x_n - y_n + z_n \qquad n = 0, 1, 2, \cdots$$
$$z_{n+1} = 2x_n + y_n + 3z_n$$

can be written in the form

$$V_{n+1} = AV_n \qquad n = 0, 1, 2, \cdots$$

where

$$V_n = \begin{pmatrix} x_n \\ y_n \\ z_n \end{pmatrix} \qquad A = \begin{pmatrix} 3 & 5 & 2 \\ 1 & -1 & 1 \\ 2 & 1 & 3 \end{pmatrix}.$$

Show further that the solution of the system can be written as

$$V_n = A^n V_0 \qquad n = 0, 1, 2, \cdots.$$

7. Let

$$P = \begin{pmatrix} 0 & 1 & 0 \\ 1 & 0 & 0 \\ 0 & 0 & 1 \end{pmatrix}.$$

---

[45] This example, and the one following, may be found in G. A. Miller, "Finite Markov Processes in Psychology," *Psychometrika*, *17* (1952), 149–167.

[46] E. B. Newman, "The Patterns of Vowels and Consonants in Various Languages," *American Journal of Psychology*, *64* (1951), 369–379.

Show[47] that if A is any 3×3 matrix, then $PA = A$ with its first and second rows interchanged whereas AP is A with its first and second columns interchanged.

8. Let[48]

$$C = \begin{pmatrix} 1.00 & -0.14 & 0.32 \\ -0.14 & 1.00 & 0.12 \\ 0.32 & 0.12 & 1.00 \end{pmatrix} \qquad C^{-1} = \begin{pmatrix} 1.15 & 0.21 & -0.40 \\ 0.21 & 1.05 & -0.20 \\ -0.40 & -0.20 & 1.15 \end{pmatrix}.$$

Calculate CC^{-1} and show that it is equal to the unit matrix I (with allowance for rounding errors.)

9. Let

$$\lambda = \begin{pmatrix} 0.61 & 0.53 & 0.76 \\ 0.27 & -0.78 & 0.53 \\ -0.75 & 0.34 & 0.38 \end{pmatrix}$$

and suppose λ' denotes the transpose of λ, i.e., the matrix obtained by interchanging the rows and columns of λ. Show[49] that $\lambda'\lambda = C$, where C is defined in the preceding problem.

10. Extend the rule for finding A^t (p. 230) to the general case of an $n \times n$ matrix with distinct characteristic roots $\lambda_1, \lambda_2, \cdots, \lambda_n$. Proceed as in (4.133) by dividing λ^t by $f(\lambda)$. Since $f(\lambda)$ is now a polynomial of degree n, the remainder r is at most of degree $(n-1)$, i.e.,

$$r(\lambda) = a_0 + a_1\lambda + a_2\lambda^2 + \cdots + a_{n-1}\lambda^{n-1}.$$

Use the Cayley-Hamilton theorem, as in the text, to show that

$$A^t = r(A) = a_0 I + a_1 A + a_2 A^2 + \cdots + a_{n-1}A^{n-1}.$$

Show that

$$\lambda_1{}^t = r(\lambda_1), \qquad \lambda_2{}^t = r(\lambda_2), \qquad \cdots, \qquad \lambda_n{}^t = r(\lambda_n).$$

This system of n simultaneous linear equations has the determinant

$$\begin{vmatrix} 1 & \lambda_1 & \lambda_1{}^2 & \cdots & \lambda_1^{n-1} \\ 1 & \lambda_2 & \lambda_2{}^2 & \cdots & \lambda_2^{n-1} \\ \cdot & \cdot & & & \cdot \\ \cdot & \cdot & & & \cdot \\ \cdot & \cdot & & & \cdot \\ 1 & \lambda_n & \lambda_n{}^2 & & \lambda_n^{n-1} \end{vmatrix}$$

[47] P is an example of a so-called elementary matrix. See Birkhoff and MacLane, footnote 35, pp. 228–233. For an application, see L. Katz, "On the Matrix Analysis of Sociometric Data," *Sociometry*, 10 (1947), 233–241. Additional material on the application of matrix algebra to the study of group structures and cliques may be found in L. Festinger, "The Analysis of Sociograms Using Matrix Algebra," *Human Relations*, 2 (1949), 153–158; F. Harary and I. C. Ross, "A Procedure for Clique Detection Using the Group Matrix," *Sociometry*, 20 (1957), 205–215.

[48] Such matrices arise in factor analysis where the matrix elements of C are correlation coefficients. This particular problem is discussed in R. B. Cattell, *Factor Analysis*, Harper, New York, 1952, p. 230.

[49] See p. 215 of Cattell, cited in footnote 48.

which is different from zero if, as assumed, all the λ's are distinct.   Hence the n equations uniquely determine the n constants $a_0, a_1, \cdots, a_{n-1}$.

*11. Extend the rule for calculating A^t (p. 230) to the case where A is a 2×2 matrix with equal characteristic roots, i.e., $\lambda_2 = \lambda_1$.   Then λ_1 is a double root of $f(\lambda) = |A - \lambda I| = 0$ so that $f(\lambda_1) = 0$ and the derivative $f'(\lambda_1) = 0$.   Differentiate (4.133) with respect to λ and show that

$$\lambda_1{}^t = r(\lambda_1) = a + b\lambda_1 \qquad t\lambda_1^{t-1} = r'(\lambda_1) = b.$$

The constants a and b are determined from these two equations rather than from (4.137).

*12. The characteristic equation $|A - \lambda I| = 0$ of matrix A in (4.157) is the equation $(\lambda - 1)^4 = 0$.   Hence A has all four of its characteristic roots equal to $\lambda = 1$.   In this case, one can prove that

$$(4.162) \qquad A^n = aI + bA + cA^2 + dA^3,$$

where the constants a, b, c, and d satisfy the simultaneous equations

$$(4.163) \qquad \begin{aligned} \lambda^n &= a + b\lambda + c\lambda^2 + d\lambda^3 \\ n\lambda^{n-1} &= \phantom{a + {}} b + 2c\lambda + 3d\lambda^2 \\ n(n-1)\lambda^{n-2} &= \phantom{a + b\lambda + {}} 2c + 6d\lambda \\ n(n-1)(n-2)\lambda^{n-3} &= \phantom{a + b\lambda + 2c + {}} 6d. \end{aligned}$$

Use this result to find A^n and thus to rederive the expression for $X(n)$ given in (4.161).   [*Hint:* From (4.157), calculate A^2 and A^3 and use (4.162) to find

$$A^n = \begin{pmatrix} a+b+c+d & 0 & 0 & 0 \\ b+2c+3d & a+b+c+d & 0 & 0 \\ c+3d & b+2c+3d & a+b+c+d & 0 \\ -b-3c-6d & -b-2c-3d & 0 & a+b+c+d \end{pmatrix}.$$

Now put $\lambda = 1$ in (4.163) and *without* solving for a, b, c, and d, use (4.163) to reduce this matrix to

$$A^n = \begin{pmatrix} 1 & 0 & 0 & 0 \\ n & 1 & 0 & 0 \\ \dfrac{n(n-1)}{2} & n & 1 & 0 \\ \dfrac{-n(n+1)}{2} & -n & 0 & 1 \end{pmatrix}.$$

Now form $A^n X(0)$ and obtain the matrix in (4.161).]

Selected
References

References to the literature are given in footnotes throughout the text. Here we list a dozen sources from the many which can profitably be consulted by a reader interested in further study of the theory and applications of difference equations.

1. Allen, R. G. D., *Mathematical Economics*, Macmillan and Company, London, 1956.
   Chapter 6 treats linear difference equations and many examples of their use in economics appear throughout the text.

2. Baumol, William J., *Economic Dynamics*, The Macmillan Company, New York, 2nd ed., 1959.
   Part IV contains an elementary treatment of difference and differential equations with applications to simple dynamic models in economics.

3. Boole, George, *A Treatise on the Calculus of Finite Differences*, G. E. Stechert, New York, 3rd ed. (reprint of 1872 2nd ed.), 1926.
   An early textbook valuable for its emphasis on the analogies between the calculus of finite differences and the differential calculus as well as for its large number of worked-out illustrative examples.

4. Fort, Tomlinson, *Finite Differences and Difference Equations in the Real Domain*, Oxford University Press, London, 1948.
   An advanced textbook.

5. Hicks, J. R., *The Trade Cycle*, Oxford University Press, London, 1950.
   Selected material on difference equations is very concisely discussed in an appendix.

6. Hildebrand, F. B., *Methods of Applied Mathematics*, Prentice-Hall, New York, 1952.
   Chapter 3 presents methods for solving difference equations and describes the application of finite difference methods to the approximate solution of problems (mostly from physics and engineering) governed by partial differential equations.

7. Jordan, Charles, *Calculus of Finite Differences*, Chelsea Publishing Company, New York, 2nd ed., 1947.
   A 654-page encyclopedic treatise with two final chapters devoted to ordinary and partial difference equations.

8. Milne, William Edmund, *Numerical Calculus*, Princeton University Press, Princeton, 1949.
   An introductory text with chapters on interpolation, finite differences, and simple difference equations.

9. Milne-Thomson, L. M., *The Calculus of Finite Differences*, Macmillan and Company, London, 1933.

An advanced, comprehensive treatise. Finite difference operators and their applications in numerical analysis are presented in the first half and the second half is devoted to difference equations.

10. Richardson, C. H., *An Introduction to the Calculus of Finite Differences*, D. Van Nostrand Company, New York, 1954.

An introductory textbook with a final chapter on difference equations.

11. Samuelson, Paul Anthony, *Foundations of Economic Analysis*, Harvard University Press, Cambridge, 1948.

Mathematical Appendix B contains a very concise, mathematically advanced treatment of difference and other functional equations.

12. Samuelson, Paul Anthony, "Dynamic Process Analysis," in F. Ellis (ed.), *A Survey of Contemporary Economics*, The Blakiston Company, Philadelphia, 1948.

A brief and elementary introduction to difference and differential equations illustrated with economic examples.

Note: Differences and indefinite sums of functions as well as solution forms for difference equations are conveniently tabulated in E. J. Cogan and R. Z. Norman, *Handbook of Calculus, Difference and Differential Equations*, Prentice-Hall, New York, 1958.

Answers
to
Problems

PROBLEMS 1.1

1 (*a*) 1, 1, 1. (*b*) 0, 0, 0, 0. (*c*) 0. **3** (*a*) 5, 8, 11. (*b*) 3, 3, 3, 3. (*c*) $3h$. **5** (*a*) 1, 2, 5. (*b*) 1, 3, 5, $2x + 1$. (*c*) $2xh + h^2$. **7** (*a*) c, $a + b + c$, $4a + 2b + c$. (*b*) $a + b$, $3a + b$, $5a + b$, $2ax + a + b$. (*c*) $2ahx + ah^2 + bh$. **9** (*a*) 1, 2, 4. (*b*) 1, 2, 4, 2^x. (*c*) $(2^h - 1)2^x$. **11** 1–4 straight lines, 5 a parabola. **13** Domain is the set of integers 0, 1, 2, $\cdots$, 7. **17** $\Delta D(p) < 0$; negative slope. **19** Put $S(p) = D(p)$ and solve to find $p = 1$; $D(1) = 1.1$; intersection is at market price.

PROBLEMS 1.2

1–4 (*a*) 0, 0. (*b*) 0. **5** (*a*) 2, 2. (*b*) $2h^2$. **7** (*a*) $2a$, $2a$. (*b*) $2ah^2$. **9** (*a*) 2, 2^x. (*b*) $(2^h - 1)2^x$. **11** $c^n y(x)$.

PROBLEMS 1.3

1 (*a*) 1. (*b*) 1. (*c*) 1. **3** (*a*) $3x + 3h + 5$. (*b*) $3x + 6h + 5$. (*c*) $3x + 3nh + 5$.

PROBLEMS 1.4

1 (*a*) $2h + 2xh + h^2$. (*b*) $-2xh - h^2$. (*c*) $5h - 10xh - 5h^2$. **2** (*a*) $15hx^{(2)}$, $30h^2x$, $30h^3$, 0, 0, $\cdots$. **9** $\Delta y(x) = 5x^4h + 10x^3h^2 + 10x^2h^3 + 5xh^4 + h^5$. **13** If $\Delta D_t = 1$ and $\Delta G_t = -4$, then $\Delta Y_t = 0$. **15** Use $\dfrac{\Delta x_t}{\Delta t} = \dfrac{\Delta x_t}{\Delta T_t} \cdot \dfrac{\Delta T_t}{\Delta t}$.

PROBLEMS 1.5

3 $n^{(k)} = n!/(n - k)!$. **5** To select groups of three: N choices for first person, $(N - 1)$ choices for second, $(N - 3)$ choices for third. Therefore $N(N - 1)(N - 2)$ ways to select group of three in which order of choice counts. Identifying as the same all groups differing only in order (e.g., 123, 132, 231, 213, 312, 321) requires dividing $N(N - 1)(N - 2)$ by 3!. **11** Put $x = 0$ in $\Delta x^m = (x + 1)^m - x^m$.

PROBLEMS 1.6

1–4 It suffices to show that difference of right-hand side equals difference of left-hand side of the equation. **7** Since $\Delta y = 0$, if Y_2 is an indefinite sum of some function, then so is $Y_1 = Y_2 + y$. Now use Theorem 1.9.

PROBLEMS 2.1

1 (a) $y_{k+1} - y_k = 0$. (c) $y_{k+2} - 2y_{k+1} + y_k = 1$. (e) $y_{k+3} - 2y_{k+2} + 2y_{k+1} = 0$.
2 (a) 1. (c) 2. (e) 2.

PROBLEMS 2.2

2 (d) $y_k = k + 1$. (f) $y_k = \dfrac{k(k-1)}{2} + 1$. (j) $y_k = \dfrac{1}{1+k}$. **3** (g) $y_k = 2^k$. (h)
$y_k = -1 + 2^{k+1} - k$. (i) $y_k = \frac{3}{2} - \frac{1}{2}(-1)^k$.

PROBLEMS 2.3

1 The coefficient of y_{k+1} is 0 at $k = 0$; if $y_0 = 0$, then y_1 may be assigned any value and there is one solution for each such assignment; if y_k with $k \neq 0$ is prescribed, then y_1 is uniquely determined, as is every other y-value. **2** When $k = 0$, the equation becomes $y_2 - y_0 = 0$, hence y_0 must be equal to y_2; if $y_0 = y_2$, then y_1 remains arbitrary; Theorem 2.1 requires that two *consecutive* y-values be prescribed.

PROBLEMS 2.4

2 (a) $y_k = \frac{1}{2}(3^{k+1} + 1)$; 2, 5, 14, 41, 122, 365; y increases without bound. (c) $y_k = 1 + (-1)^k$; 2, 0, 2, 0, 2, 0; y oscillates between the values 2 and 0. (e) $y_k = 1 + (-\frac{1}{2})^k$; 2, $\frac{1}{2}$, $\frac{5}{4}$, $\frac{7}{8}$, $\frac{17}{16}$, $\frac{31}{32}$; y approaches the value 1 but is alternately above and below 1. **3** (a) $y_k = \frac{11}{2} \cdot 3^k + \frac{1}{2}$. (c) $y_k = 1 + 5(-1)^k$. (e) $y_k = 1 + 5(-\frac{1}{2})^k$; behavior of y-values does not change. **5** $y_k = 3(2^{k-1} - 1)$, $k = 1, 2, \cdots, 6$.

PROBLEMS 2.5

1 (a) C3. (b) D1. (c) C3. (d) C4. (e) D4. (f) C2. (g) D2. (h) C4. (i) D3. (j) D4. **2** (a), (c), (d), (f) (h), and (i) are bounded, the others unbounded; sequence (i) is bounded but not convergent.

PROBLEMS 2.6

1 (a) $y_k = \frac{1}{2}$, row (a) in Table 2.2. (b) $y_k = \frac{1}{2}(3^k + 1)$, row (b) in Table 2.2. (c) $y_k = \frac{1}{4}[5(-3)^k - 1]$, row (h) in Table 2.2. (d) $y_k = 2[(\frac{1}{2})^k + 1]$, row (d) in Table 2.2. (e) $y_k = \frac{1}{5}[4(-\frac{2}{3})^k + 1]$, row (f) in Table 2.2. (f) $y_k = 5 - k$, row (k) in Table 2.2. (g) $y_k = (-1)^{k+1}$, row (g) in Table 2.2.

PROBLEMS 2.7

1 (a) 50 years. (b) 35 years. **4** (a) \$2687.04. (b) $R = \$186.08$. **6** Outstanding principals at beginning of year 1, 2, 3, $\cdots$ are \$100.00, 92.05, 83.70, 74.93, 65.73, 56.07, 45.92, 35.27, 24.08, 12.33, 0. **10** (a) \$7.95. (b) \$7.59. **11** Must accumulate \$100 in 10 years so annual payment is \$7.95. Book values after first, fifth, and last payments are \$112.05, \$76.07, and \$20.03 respectively. **12** If debt is S and interest rate i, then periodic payment for interest and sinking fund is $Si + S\dfrac{1}{s_{\overline{n}|i}}$ and for amortization is $S\dfrac{1}{a_{\overline{n}|i}}$. Now use the identity in Problem 9. **13** Note that A_{k+1}
$= \dfrac{R}{i}[(1 + i)s_{\overline{k+1}|i} - (k + 1)]$ and use the identity $s_{\overline{k+1}|i} = s_{\overline{k}|i}(1 + i) + 1$.

PROBLEMS 2.8

1 If $Y_0 = Y^*$, then $Y_t = Y^*$ (for all t) by (2.83); $\Delta Y_t = 0$ so $I_t = 0$ by (2.77); $C_t = Y^*$ follows from (2.74). **3** (i) $S_t = sY_t$; (ii) $I_t = g(Y_t - Y_{t-1})$; (iii) $S_t = I_t$. Put (i) and (ii) in (iii) and simplify to find the difference equation; $Y_t = Y_0 c^t$; under given conditions, $c > 1$ so $\{Y_t\}$ steadily diverges to $+\infty$. **5** (i) becomes $S_t = sY_{t-1}$; if $s > 0$ and $g > 0$, then $k > 1$. **7** $S_t = I_t$ is now $sY_t = g(Y_t - Y_{t-1}) + KY_t + L$; use Theorem 2.7; use the difference equation of Problem 3 to show $\dfrac{Y_{t+1} - Y_t}{Y_{t+1}} = \dfrac{c-1}{c}$

$= \dfrac{s}{g}$; write $S_{t+1} = I_{t+1}$ or $sY_{t+1} = g(Y_{t+1} - Y_t) + KY_{t+1} + L$, transpose $KY_{t+1} + L$,

and then divide both sides by gY_{t+1} to obtain $\dfrac{Y_{t+1} - Y_t}{Y_{t+1}} = \dfrac{(s-K)}{g} - \dfrac{L}{gY_{t+1}}$; since the subtracted amount gets smaller as Y_{t+1} increases, the whole quantity increases. **9** (a) $p_{t+1} = 2 - q_{t+1} = 2 - (p_{t+1}^* + 1) = 1 - p_{t+1}^*$ so $p_{t+1} = p_{t+1}^*$ only if each is equal to $\frac{1}{2}$. (b) Use result of (a) to write $p_{t+1} = 1 - \alpha P - (1 - \alpha)p_t$, put $p_{t+1} = P$, and solve for P.

PROBLEMS 2.9

1 The equation involves y_{t-2} and y is defined only for nonnegative t-values. **2** (a) $y_t = -25(0.2)^t + 150$; with headings as in Table 2.4, first three rows of table are 0, –, 125, 25, 100; 1, 25, 145, 29, 96; 2, 29, 149, 29.8, 95.2. (b) $y_t = -100(0.8)^t + 600$; first three rows of table are 0, –, 500, 400, 100; 1, 400, 520, 416, 84; 2, 416, 536, 428.8, 71.2; convergence toward limiting values is slower as β increases. **3** Inventory level at end of period t (i_t) = inventory level at end of previous period (i_{t-1}) + goods produced for inventory (s_t) + [goods produced for sale (u_t) – goods sold (βy_t)]. **5** In (a) use $i_0 = 100$, $i_1 = 96$, $\beta = 0.2$ to get $c = 76$ and $i_t = 5(0.2)^t + 95$; in (b) use $i_0 = 100$, $i_1 = 84$, $\beta = 0.8$ to get $c = 4$ and $i_t = 80(0.8)^t + 20$; in both cases, substituting $t = 2, 3, \cdots$ will yield values of i_t which should match those in the tables of Problem 2.

PROBLEMS 2.10

3 (a) $p_{n+1} = 0.5p_n + 0.1$ has the solution (by Theorem 2.7) $p_n = (0.3)(0.5)^n + 0.2$ so that $p_n \to 0.2$; i.e., the limiting probabilities of response and nonresponse are 0.2 and 0.8 respectively, or in ratio $1 : 4$. (b) From solution (2.105), $p_n \to \dfrac{a}{a+b}$; i.e., the limiting probabilities of response and nonresponse are $\dfrac{a}{a+b}$ and $\dfrac{b}{a+b}$ respectively, or in ratio $a : b$. **5** p_n increases steadily and approaches 1 in each case. **7** If $a = 0$ and $b = 1$, then, independently of p_0, (2.105) shows that $p_n = 0$ for all n (one-trial extinction). **9** $p_k = Ep_{k-1} = (1 - b)p_{k-1} = (1 - b)Ep_{k-2} = (1 - b)^2 p_{k-2} = \cdots = (1 - b)^{k-1}p_1 = (1 - b)^{k-1}Qp_0 = (1 - b)^{k-1}[p_0 + a(1 - p_0) - bp_0]$. **11** Use Theorem 2.7 with $A = \left(1 - \dfrac{s_c}{S_c}\right)$, $B = s_c$. **13** Note that the limiting probability of an individual answering "yes" does not depend on $f(0)$, his initial probability of answering "yes." **15** Write the equation in standard form $A_{n+1} = (1 - p)A_n + i(1 - p)$ and use Theorem 2.7.

PROBLEMS 2.11

1 $w_k = 2^k$, $U_k = (0.5)k$. **3** Changing w_k to S_k and U_k to R_k reduces (2.115) to (2.116). **4** (a) $p_0 = a$. (b) $p_{k+1} = ar^{k+1} = rp_k$. (c) Solution for p is of form (2.114) with $\alpha = r - 1 < 0$. **5** Relative increase is given by $\Delta N_t/N_t$; solution by Theorem 2.7 with $A = 1.02$, $B = 0$; $t = 35$ years required for doubling. **7** $N_{t+1} = (1.02)N_t + 10$; solution by Theorem 2.7 with $A = 1.02$, $B = 10$; if $N_t = 2N_0$, then $(1.02)^t = \dfrac{2N_0 + 500}{N_0 + 500}$

and number of years required for doubling does depend upon N_0. **9** Constant rate of growth is interpreted as $\Delta W_t = cW_t$, and $c = 0.05$ here.

PROBLEMS 3.1

1 (a) $a_1 = 3$, $r_k = 8$. (b) $a_1 = -1$, $r_k = 1$. (c) $a_1 = -\frac{5}{2}$, $r_k = \frac{1}{2}(3k + 1)$. (d) $a_1 = 0$, $a_2 = -1$, $r_k = k^2$. **3** In each case, show that the equation is satisfied for all k if y_k is replaced by $y_k{}^*$. **4** (a) $y_k = C(-3)^k + 2$, (b) $y_k = C + k$, (c) $y_k = C(\frac{5}{2})^k - k - 1$; for particular solutions put (a) $C = 2$, (b) $C = 4$, (c) $C = 5$. **5** Model proofs on that of Theorem 3.1. **7** In asserting $(-a_1)^0 = 1$.

PROBLEMS 3.2

1 $y_{k+2}^{(1)} = y_k^{(1)} = 1$ and $y_{k+2}^{(2)} = (-1)^{k+2} = (-1)^k(-1)^2 = (-1)^k = y_k^{(2)}$ so $y^{(1)}$ and $y^{(2)}$ are both solutions. They form a fundamental set since determinant (3.26) equals $-2 \neq 0$. The general solution is given by $y_k = C_1 + C_2(-1)^k$, the particular solution is $y_k = 1 - (-1)^k$. **3** $y_k = C_1 + C_2(-1)^k + \frac{1}{6}(k^3 - 3k^2 + 2k)$; the particular solution is obtained by choosing $C_1 = 1$, $C_2 = -1$.

PROBLEMS 3.3

1 (a) $Y_k = C_1 + C_2(-1)^k$. (c) $Y_k = (C_1 + C_2k)(-1)^k$. (e) Roots: $1 \pm \dfrac{\sqrt{3}}{3}i$
$= \dfrac{2}{\sqrt{3}}\left(\cos\dfrac{\pi}{6} \pm i\sin\dfrac{\pi}{6}\right), y_k = A\left(\dfrac{2}{\sqrt{3}}\right)^k \cos\left(\dfrac{k\pi}{6} + B\right)$. **2** (a) $y_k = \frac{1}{2}[1 - (-1)^k]$. (c)
$y_k = k(-1)^{k+1}$. (e) $y_k = \sqrt{3}\left(\dfrac{2}{\sqrt{3}}\right)^k \sin\dfrac{k\pi}{6}$. **3** Using the notation of Section 2.5,

sequences are of type (a) D3, (b) D1, (c) D4, (d) C3, (e) D4, (f) D4.

PROBLEMS 3.4

1 (a) Let $y_k{}^* = A$ and show that $A = 1$. (c) $y_k{}^* = A$ fails; let $y_k{}^* = Ak$ and find $A = -1$. (e) $y_k{}^* = A\sin\dfrac{k\pi}{2} + B\cos\dfrac{k\pi}{2}$ fails; let $y_k{}^* = Ak\sin\dfrac{k\pi}{2} + Bk\cos\dfrac{k\pi}{2}$ and find $A = -\frac{1}{2}$, $B = 0$. (g) Try $y_k{}^* = Ak^2 + Bk^3$ and find $A = 1$, $B = \frac{1}{2}$. **2** To the particular solutions obtained in Problem 1 add the general solution Y of the homogeneous equation, where Y_k equals (a) $C_1 2^k + C_2 3^k$, (c) $C_1 + C_2 2^k$, (e) $A\cos\left(\dfrac{k\pi}{2} + B\right)$, (g) $C_1 + C_2 k$, (h) $C_1 + C_2 k$. **3** (a) $y_k = 2^{k+1} - 2 \cdot 3^k + 1$. (c) $y_k = 2 - 2^k - k$. (e) $y_k = \dfrac{\sqrt{5}}{2}\cos\left(\dfrac{k\pi}{2} + B\right) - \dfrac{k}{2}\sin\dfrac{k\pi}{2}$, where B is the acute angle for which $\sin B = \dfrac{1}{\sqrt{5}}$

and $\cos B = \dfrac{2}{\sqrt{5}}$. (g) $y_k = \frac{1}{2}(2 - 7k + 2k^2 + k^3)$.

PROBLEMS 3.5

1 (a) Oscillates finitely, $\rho = 1$. (c) Oscillates infinitely, $\rho = 1$. (e) Oscillates infinitely, $\rho = \dfrac{2}{\sqrt{3}} > 1$. **2** (a) (i) $y_k = 2^{k+2} - 3^{k+1}$; (ii) diverges to $-\infty$; (iii) diverges to $-\infty$. (c) (i) $y_k = 3 - 2^{k+1}$; (ii) diverges to $-\infty$; (iii) diverges to $-\infty$. (d) (i) $y_k = (-1)^k$; (ii) oscillates finitely; (iii) diverges to $-\infty$. (e) (i) $y_k = \sqrt{2}\cos\left(\dfrac{k\pi}{2} + \dfrac{\pi}{4}\right)$;

(ii) oscillates infinitely; (iii) oscillates infinitely. (g) (i) $y_k = 1 - 2k$; (ii) diverges to $-\infty$; (iii) diverges to $+\infty$. (h) (i) $y_k = 1 - 2k$; (ii) diverges to $-\infty$; (iii) diverges to $+\infty$. **3** $\rho = r = \sqrt{\alpha}$.

PROBLEMS 3.6

1 $Y_{t+2} - \frac{1}{2}Y_{t+1} = 1$, $Y_t = (\frac{1}{2})^{t-1} + 2$ ($t = 1, 2, \cdots$), $\{Y_t\}$ steadily decreases to limit 2. **3** Same as derivation of (3.75) in text. **5** $N_t = C_1(\sqrt{2})^t + C_2(-\sqrt{2})^t$ and using initial conditions, find $C_1 = \frac{1}{2} + \frac{1}{2\sqrt{2}}$, $C_2 = \frac{1}{2} - \frac{1}{2\sqrt{2}}$; simplifying the solution shows that

$N_t = (\sqrt{2})^t$ if t is even, $N_t = (\sqrt{2})^{t-1}$ if t is odd; since $(\sqrt{2})^t = 2^{\frac{t}{2}}$ and $(\sqrt{2})^{t-1} = 2^{\frac{t-1}{2}}$, we calculate $\log_2 N_t = \frac{t}{2}$ if t is even and $\log_2 N_t = \frac{t-1}{2}$ if t is odd, hence $C = \frac{1}{2}$.

6 Auxiliary equation $m^2 - m + \frac{1}{2} = 0$ has roots $= \frac{1}{\sqrt{2}}\left(\cos\frac{\pi}{4} \pm i\sin\frac{\pi}{4}\right)$; solution is

$y_t = A\left(\frac{1}{\sqrt{2}}\right)^t \cos\left(\frac{t\pi}{4} + B\right) + 1000$; find $A = -200$, $B = \frac{\pi}{2}$; to reduce to required

form note that $\cos\left(x + \frac{\pi}{2}\right) = -\sin x$; $y_2 = 1100$, $y_3 = 1050$, $y_4 = 1000$, $y_5 = 975$, $y_6 = 975$, $y_7 = 987.5$, $y_8 = 1000$. **7** $i_1 = 550$, $i_2 = 600$, $i_3 = 625$, $i_4 = 625$, $i_5 = 612.5$, $i_6 = 600$, $i_7 = 593.75$, $i_8 = 593.75$; although income stops rising after $t = 2$, inventory continues to increase up to $t = 4$; y_t starts rising again at $t = 6$ but i_t does not increase until $t = 8$. **9** Anticipated sales in period t equal 120 (if $\eta = 0$), 130 (if $\eta = \frac{1}{2}$), 140 (if $\eta = 1$), 110 (if $\eta = -\frac{1}{2}$), 100 (if $\eta = -1$). **11** (a) $p_t = A\cos\left(\frac{t\pi}{3} + B\right) + Kr_0$; oscillates finitely. (b) $p_t = A(\sqrt{2})^t \cos\left(\frac{t\pi}{4} + B\right) + Kr_0$; oscillates infinitely. (c)

$p_t = A\left(\frac{1}{\sqrt{2}}\right)^t \cos(t\theta + B) + Kr_0$, where θ is angle for which $\sin\theta = \frac{\sqrt{14}}{4}$,

$\cos\theta = \frac{\sqrt{2}}{4}$; damped oscillations around limit Kr_0. **13** Auxiliary equation is $m^2 - (2 - A)m + (1 - A + B) = 0$; find roots by quadratic formula; solution is $P_n = C_1 r_1^n + C_2 r_2^n$ and determine C_1 and C_2 by using prescribed values P_0 and P_1.

PROBLEMS 3.7

1 (a) $y_k = C_1 + C_2 2^k + C_3 3^k$. (c) Roots: $1, 1 + i, 1 - i$, $y_k = A(\sqrt{2})^k \cos\left(\frac{k\pi}{4} + B\right)$ $+ C$. (d) Roots: $1, 1, i, -i$, $y_k = C_1 + C_2 k + A\cos\left(\frac{k\pi}{2} + B\right)$. (e) $y_k = C_1 + C_2 k + C_3 k^2 + C_4 k^3$. **2** Auxiliary equation $m^{n+1} - (1 + A)m^n + A = 0$; use long division to divide by $m - 1$; when $m = 1$, bracketed quantity is $1 - nA < 0$ since $A > \frac{1}{n}$; when $m \neq 1$, by summing geometric progression, write bracketed quantity as $m^n \dfrac{1 - m^n}{1 - m} = m^n\left(1 - \dfrac{A}{m - 1}\right) + \dfrac{A}{m - 1}$ which is positive if $m - 1 > A$.

PROBLEMS 4.1

1 Let $y_k^* = \dfrac{B}{1-A}$. Stability requires the solution y (given in Theorem 2.7) to approach y^* for all y_0 and this occurs (see Table 2.2) if and only if $-1 < A < 1$. **3** $a_1 = -B$ and $a_2 = C$ in (4.13); the first two inequalities, when solved for B, yield $B < 1 + C$ and $B > -1 - C$, and the third inequality is $C < 1$. **5** (a) In (4.13) put $a_1 = -c_1$ and $a_2 = -c_2$. The first inequality in (4.13) becomes $1 - c_1 - c_2 > 0$ and is proved as follows: $1 - c_1 - c_2 = 1 - (c_1 + c_2) > 1 - (|c_1| + |c_2|) > 1 - 1 = 0$; the other two inequalities in (4.13) also follow from the hypothesis. (b) Put $y_n^* = \alpha + \beta n$ for y_n in (4.19) and equate constant terms and coefficients of n on both sides to obtain two linear equations for the undetermined coefficients α and β. The second equation is $\beta(1 - c_1 - c_2) = h$ from which β may be found if $1 - c_1 - c_2 \neq 0$. The constant α may likewise be determined if $c_1 + c_2 \neq 1$.

PROBLEMS 4.2

1 (a) $y_k = Ck!$; put $C = 1$. (b) $y_k = k!(2^k + c)$; put $c = 0$. (c) $y_k = C3^0 \cdot 3^1 \cdot 3^2 \cdots 3^{k-1} = C3^{1+2+\cdots+(k-1)} = C3^{\frac{k(k-1)}{2}}$; put $C = 1$. (d) $y_k = C(-1)^{\frac{k(k-1)}{2}}$; put $C = 1$. **5** (a) $p_{t+1} = -\frac{1}{2}p_t + \frac{1}{2}$, $p^* = \frac{1}{3}$, stable. (b) $p_{t+1} = -p_t + 4$, $p^* = 2$, unstable. (c) $p_{t+1} = -2p_t + 6$, $p^* = 2$, unstable.

PROBLEMS 4.3

1 Use $\Delta^2 y_k = y_{k+2} - 2y_{k+1} + y_k$ and $Ey_k = y_{k+1}$. **3** Use the results of Theorem 4.3, with $N = 2$, to obtain $\alpha_1 = 1$, $\alpha_2 = 3$, $y_k^{(1)} = \sin\dfrac{k\pi}{3}$, $y_k^{(2)} = \sin\dfrac{2k\pi}{3}$. **4** Characteristic functions are given by $y_k^{(n)} = (\sqrt{\alpha_n})^k \sin\dfrac{n\pi k}{N+1}$ $(n = 1, 2, \cdots, N)$ and corresponding characteristic values are $\alpha_n = \frac{1}{4}\sec^2\dfrac{n\pi}{N+1}$ $(n = 1, 2, \cdots, N)$. All nontrivial solutions are obtained when $\frac{1}{4} < \alpha$.

PROBLEMS 4.4

1 (a) $\dfrac{2}{1-s} + \dfrac{3s}{(1-s)^2}$. (b) $\dfrac{q}{1-ps}$. (c) $\dfrac{s^2}{1-s}$. (d) Use lines (3) and (7) of Tables 4.1 and 4.2 to get $\dfrac{2}{(1-s)^3}$. (e) $\dfrac{s^2}{(1-s)^3}$. (f) s. **3** (a) $y_k = 3 + 2^k$. (b) $y_0 = 0$, $y_k = k + 1$ if $k > 0$. (c) $y_0 = y_1 = 0$, $y_k = 1$ if $k > 1$. (d) $y_0 = 0$, $y_k = 2^{k-1}$ if $k > 0$. (e) $y_0 = 1$, $y_1 = 4$, $y_2 = 6$, $y_3 = 4$, $y_4 = 1$, $y_k = 0$ if $k > 4$. (f) $y_k = \binom{n}{k}(-1)^k 3^{n-k}$ if $k = 0, 1, \cdots, n$; $y_k = 0$ if $k > n$. **5** Apply the generating function transformation and simplify to find

$$Y(s) = \frac{5s^2 - 4s + 1}{(1-s)(1-2s)(1-3s)} = \frac{1}{1-s} + \frac{-1}{1-2s} + \frac{1}{1-3s}.$$

Apply an inverse transformation to get $y_k = 1 - 2^k + 3^k$. **6** (a) 0.0064. (b) 0.0067 **7** The sum in (4.55) is equal to $Y_n(1)$ which in turn is equal to 1 since $p + q = 1$. **9** (a) $P(X = k) =$ probability that the specified success is followed by $(k - 1)$ consecutive successes and then a failure $= p^{k-1}q$. (b) Use result of Problem 8 to get

$E(X) = 2$. **11** (a) npq. (b) p/q^2. **13** $\underset{1\ 1}{\Delta}(\Delta z) = z_{j+2,k} - 2z_{j+1k} + z_{j,k}$, $\underset{2\ 2}{\Delta}(\Delta z) = z_{j,k+2}$

$- 2z_{j,k+1} + z_{j,k}$, $\underset{1\ 2}{\Delta}(\Delta z) = \underset{2\ 1}{\Delta}(\Delta z) = z_{j+1,k+1} - z_{j+1,k} - z_{j,k+1} + z_{j,k}$. **15** Note that

$\underset{1}{\Delta}(\tau_k p_{k,n}) = \tau_{k+1} p_{k+1,n} - \tau_k p_{k,n}$. **17** The product contains k factors each containing s,

so P_k has the form $p_{k,k}s^k + p_{k,k+1}s^{k+1} + \cdots$. The probabilities $p_{k,n}$ with $n < k$
are the coefficients of s^n with $n < k$ and must therefore be 0. **19** If X_n denotes the number of times a word is recalled in the first n trials, then X_n assumes the value k with
probability $p_{k,n}$ ($k = 0, 1, 2, \cdots, n$) and the formula for $E_n = E(X_n)$ follows from
(4.63). The difference $X_{n+1} - X_n$ is the number of times a word is recalled on trial
number $n + 1$. This difference is either 0 or 1 and its expected value is the probability
that the difference is 1 or, equivalently, the proportion of words recalled on trial number
$n + 1$. Hence $\rho_{n+1} = E(X_{n+1} - X_n) = E_{n+1} - E_n$. See cited reference for proof of
equation for ρ_{n+1}.

PROBLEMS 4.5

1 (a) $A + B = \begin{pmatrix} 0 & 3 \\ 4 & 2 \end{pmatrix}$; $B + D = \begin{pmatrix} 1 & 1 \\ 1 & 3 \end{pmatrix}$, $A + C$ undefined. (b) $AB = \begin{pmatrix} 1 & 1 \\ -1 & 3 \end{pmatrix}$,

$BA = \begin{pmatrix} 2 & 0 \\ 1 & 2 \end{pmatrix}$. (d) $AC = \begin{pmatrix} 11 \\ 13 \end{pmatrix}$, CA undefined. (f) In general, $AB \neq BA$. Hence
one cannot simplify by writing $2AB$ in place of the sum $AB + BA$. (g) $|A| = -4$
$\neq 0$. (i) $|A - \lambda I| = (1 - \lambda)(2 - \lambda) - 6 = (\lambda + 1)(\lambda - 4)$. Characteristic vectors

associated with $\lambda_1 = -1$ and $\lambda_2 = 4$ are $C_1 = \begin{pmatrix} 1 \\ -1 \end{pmatrix}$ and $C_2 = \begin{pmatrix} 2 \\ 3 \end{pmatrix}$ respectively.

(j) $D = \begin{pmatrix} -1 & 0 \\ 0 & 4 \end{pmatrix}$, $M = (C_1 \ \ C_2) = \begin{pmatrix} 1 & 2 \\ -1 & 3 \end{pmatrix}$. (o) 2 and 3. **4** (a) State 1 = right
response given, state 2 = wrong response; $\alpha = 0.03$, $\beta = 0.27$; p_t = probability of
right response in experiment number t; use (4.141) to find p_t; $p_t \to 0.9$ as $t \to \infty$.
(b) State 1 = vowel, state 2 = consonant; $\alpha = 0.49$, $\beta = 1$; p_t = probability that letter
number t is a vowel; $p_t \to 0.67$ as $t \to \infty$. (c) State 1 = good weather, state 2 = bad
weather; $\alpha = \beta = 0.1$; p_t = probability that day number t following some prescribed
initial day has good weather; $p_t \to \frac{1}{2}$. **5** (a) $x_t = \frac{1}{2}(3^t - 1)$. (b) $x_t = \frac{1}{3}[2^t + 2(-1)^t]$.
(c) $x_t = 2 - (-1)^t$.

Index

DATE DUE

MAY 8 ' 79			
AUG M1 '80			
AP 9 '93			
ME 4 '93			